GIS Data Conversion Handbook

GIS Data Conversion Handbook

Glenn E. Montgomery
Harold C. Schuch

UGC Consulting

1993
GIS World Books

GIS World, Inc.
Fort Collins, Colorado, USA

Publisher:	H. Dennison Parker
Vice President, Publishing:	Derry Eynon
Illustrator:	Julie Laugen
Production Manager:	Christine Thompson
Cover Design and Graphics/Composition:	Darin E. Sanders Wade L. Smith
Printer:	Edwards Brothers, Inc.

Copyright © 1993 GIS World, Inc. and UGC Consulting.

All rights reserved. No part of this publication may be reproduced, stored in a retrieval system, or transmitted, in any form or by any means, electronic, mechanical, photocopying, recording, or otherwise, without the prior written permission of UGC Consulting and GIS World, Inc. Printed in the United States of America.

Library of Congress Cataloging-in-Publication Data

Montgomery, Glenn E., 1951-
    GIS data conversion handbook / Glenn E. Montgomery, Harold C. Schuch.
    320 p. 19 x 23 cm.
    Includes index.
    ISBN 0-9625063-4-6
      1. Geographic information systems.   2. Data base management.
  I. Schuch, Harold C., 1943-   . II. Title.
G70.2.M66   1993
910' .285—dc20   92-28346
                                                                                            CIP

ISBN 0-9625063-4-6

ABCDEFGHIJKLMNOPQST

GIS World Books published by: GIS World, Inc.
                                                          155 E. Boardwalk Drive
                                                          Suite 250
                                                          Fort Collins, CO 80525, USA

Chapters-at-a-Glance

Chapter 1 AM/FM/GIS and Their Markets
The reader is introduced to automated mapping/facilities management (AM/FM) and geographic information systems (GIS) through historic perspectives and comparisons, the identification of GIS benefits, and a description of the market sectors that already are benefiting from this technology.

Chapter 2 GIS Data Conversion
Data conversion is described briefly in terms of conversion requirements, data sources, conversion techniques, and aspects of conversion management.

Chapter 3 Hardware Issues
In general terms, hardware evolution in relation to GIS is presented in a historic perspective that includes a modern configuration with its components and peripherals.

Chapter 4 GIS Data Sources
A variety of document and information sources usually are converted into a GIS database. The principal sources are presented and discussed. Source quality, content, and format also are reviewed.

Chapter 5 Land Base Data vs. Facilities Data
The conversion of land base data usually has different requirements than the conversion of facilities data. These data types are discussed and compared.

Chapter 6 Data Conversion Methods
Data conversion methods are identified and described. Conversion methods usually are used in various combinations. These and supporting technologies are presented.

Chapter 7 GIS Data Quality
Databases for GIS must satisfy certain requirements that vary from one project to another. These requirements for data conversion are introduced.

Chapter 8 GIS Database Design
Database design is a necessary prerequisite to the population of a GIS database. The design stages and elements are presented, together with a brief description of the tools that allow database structuring.

Chapter 9 Multiparticipant Databases
A sound way to reduce the cost of GIS database acquisition is to share the conversion costs with others. This chapter examines this increasingly attractive conversion solution which may harbor difficulties that are not readily anticipated.

Chapter 10 Outsourcing GIS Data Conversion
Data conversion services are available through a variety of service providers. Criteria for selecting a conversion service provider are discussed.

Chapter 11 Impact of Emerging Technologies
Data conversion can be costly and time-consuming. As presented in this chapter, emerging data conversion technologies try to reduce cost and time, usually by automating portions or all of the process.

Chapter 12 Developing a GIS Data Conversion Plan
This chapter reviews the major components of a conversion plan and puts them into perspective, providing a general overview of the conversion process.

Contents

Preface	xxi
Acknowledgments	xxiii
List of Figures	xxv
List of Tables	xxvii

1   AM/FM/GIS and Their Markets   1

1.1	Introduction	1
1.2	Origins of GIS Technology	2
	Computer-Aided Drafting and Design (CADD or CAD)	5
	Database Management Systems (DBMS)	6
	Automated Mapping (AM)	8
	Facilities Management (FM)	9
	Automated Mapping/Facilities Management (AM/FM)	10
	Geographic Information Systems (GIS)	11
1.3	Potential Benefits of a GIS	12
	Quantitative Benefits	12
	Qualitative Benefits	13
1.4	Typical Database Information Layers	14
1.5	GIS Market Sectors	17
1.6	Public Sector	18
	Federal Government	18
	State Government	19
	Local Government	20
1.7	Regulated Sector	21
	Electric Utilities	21
	Gas Utilities	22
	Telephone Companies	22
1.8	Private Sector	23
	Natural Resource Companies	23

	Timber and Forest Products Companies	24
	Financial Services and Insurance Companies	24
	Real Estate Companies	25
	Transportation Companies	25
	Retail Companies	25

2 GIS Data Conversion — 27

2.1	Introduction	27
2.2	GIS Data Conversion Issues	28
	GIS Data Conversion Effort	31
	Developing the Data Conversion Approach	32
2.3	GIS Data Requirements of the System	34
	Data Relationships	35
	GIS Vendors and GIS Data Formats	35
2.4	GIS Source Data	35
2.5	Techniques Used to Populate a GIS Database	37
2.6	Making It All Fit Together	39
	Conversion Management	40
	Risk	44
	Information Services and Data Conversion	44
	Corporate View of Data	45

3 Hardware Issues — 47

3.1	Introduction	47
3.2	Factors Influencing Data Conversion Hardware	47
	The Need to Convert Certain Data	48
	Available Technology	48
	Economics Dictates	48
	Industry Experience	48

3.3	Data Conversion Hardware Evolution	49
	Prior to 1980	49
	The 1980s	49
	The 1990s	51
3.4	In-House (Internal) vs. External Conversion Hardware Issues	53
3.5	Input Hardware	55
	Keyboard	55
	Command Tablet	55
	Graphic Input Device	55
	Digitizing Tablets	56
	Scanners	56
	Pen-Based Laptop/Hand-Held PCs	56
3.6	Output Hardware	56
	Pen Plotters	56
	Raster Plotters	57
	Screen Copy Devices	57
	Computer FAX	58
	Printers	58
3.7	GIS Data Conversion Workstations	58
	Digitizing Workstations	59
	Review/Edit Workstations	59
	Review/Tabular Attribute Data Input Workstations	59
	X Terminals	60
3.8	Storage Devices	60
	Magnetic Disk	60
	Optical Disk	61
	Magnetic Tape	62
3.9	Processors	62
	Workstations	63
	Servers	63
	Minicomputers	63
	Mainframe Computers	64
3.10	Communications	64

3.11	Survey Input	65
	Photogrammetric Devices	65
	Total Stations	65
	GPS	66
3.12	Field Data Entry Stations	66

4 GIS Data Sources 67

4.1	Introduction	67
4.2	Key Issues Involving Data Sources	67
	Accuracy	68
	Coverage	68
	Completeness	69
	Timeliness	69
	Correctness	70
	Credibility	70
	Validity	70
	Reliability	71
	Convenience	71
	Condition	71
	Readability	71
	Precedence	72
	Maintainability	72
4.3	Maps	72
	Users of Maps	74
	Range of Scale	74
	Typical Sheet Size	75
	Media	79
	Typical Content	79
	Importance of Symbology	80
	Tools Used in Original Map Production	80

4.4	Drawings	82
	Users of Drawings	82
	Range of Scale	83
	Typical Sheet Size	83
	Media	84
	Importance of Symbology	84
	Tools Used in Original Drawing Production	84
4.5	Aerial Photographs	84
	Decision to Use	85
	Typical Scales	86
	Color vs. Black and White	87
	Typical Features Captured through Aerial Mapping	87
	Related Intermediate Products	88
4.6	Cards and Records	88
4.7	Existing Databases	89
	Community-Oriented Data	90
	Accounting-Oriented Data	90
	Assessment-Oriented Data	91
	Engineering-Oriented Data	91
	Commercial Databases	91
	CAD Graphic Database	93

5 Land Base Data vs. Facilities Data — 95

5.1	Introduction	95
5.2	Land Base Data	97
	Planimetric Data	97
	Cadastral Data	98
	Hypsographic Data	98
	Control Data	99
	Administrative Data	99
5.3	Facilities Data	99
5.4	Land Base vs. Facilities Base Issues	102

6 Data Conversion Methods — 105

6.1	Introduction	105
6.2	Map Digitizing	105
	Equipment	106
	Setup	107
	Digitizing	108
	Advantages and Disadvantages	109
6.3	Keyboard Entry	109
	Attribute Database	110
	Graphics Database	111
	Precise Calculation	111
6.4	Photogrammetry	113
	Photogrammetry Process Overview	114
	Why Use Photogrammetry?	114
	Ground Control	116
	Photography	116
	Aerial Triangulation	117
	Stereoplotters	117
	Digitizing	117
	Orthophotography	118
	Future Advancements	119
6.5	Scanning	119
	Scanning a Document	121
	Image Storage	121
	Uses for Scanned Images	122
6.6	Automated Conversion	123
	Lines and Curves	123
	Symbols	124
	Text	124
	Batch vs. On-Line Conversion	124
6.7	Field Survey	125
	Conventional Field Survey	125
	GPS Surveys	126

6.8	Field Inventory	128
	Data Collection	129
	Data Verification	130
6.9	Data Translation	130

7 GIS Data Quality   131

7.1	Introduction	131
7.2	Cartographic Quality	131
	Relative Accuracy	132
	Absolute Accuracy	132
	Graphic Quality	133
7.3	Informational Quality	134
	Completeness	134
	Correctness	135
	Timeliness	135
	Integrity	135
7.4	Measuring GIS Positional Accuracy	136
	Accuracy and Scale	136
	Positional Accuracy Measurements	139
	Reproduction, Scale, and Accuracy	140
7.5	Typical GIS Positional Accuracy Requirements	140
7.6	Why Positional Accuracy Is an Issue	141
7.7	Verifying GIS Data Quality	143
	Automated QA	143
	Manual QA	144

8 GIS Database Design   147

8.1	Introduction	147

8.2	Database Design Stages	147
	Conceptual Database Design	148
	Physical Database Design	149
	Database Implementation	149
8.3	What Drives Database Design?	149
	User Needs	150
	GIS Application Data Requirements	151
	Available Data and Cost of Conversion	151
	Data Conversion Schedule	153
	Future Expansion	153
	Maintainability	153
8.4	Database Design Elements	154
	Logic Elements	154
	Graphic Elements	156
	Attributes	156
	GIS Data Relationships	157
	Existing Digital Data	158
	Raster Image Data	158
	Other Data Formats	158
8.5	Graphic Structure	159
	Symbology	159
	Color	160
	Geometric Integrity	160
	Text Annotation	161
	Layers	161
	Visibility Rules	162
8.6	Topologic Structure	162
	Node Topology	163
	Line Topology	163
	Area Topology	163
	Attribute Topology	164
8.7	Tabular Structure	164
	Flat Files	164
	Hierarchical Model	164
	Relational Model	165

| 8.8 | Interfaces | 166 |

9 Multiparticipant Databases — 167

9.1	Introduction	167
9.2	Standards	169
9.3	Organization	170
	The Executive Steering Committee	170
	The Project Champion	170
	The Technical Project Team	171
	The Project Manager	171
9.4	Agreements	171
9.5	Funding	172
9.6	Data Ownership	174
9.7	Liability	174
9.8	Risk	176
9.9	Feasibility Study	178
9.10	Database Contents	178
9.11	Maintenance	178

10 Outsourcing GIS Data Conversion — 181

10.1	Introduction	181
	Evolution	181
	Services	183
	External vs. Internal Conversion	184
10.2	Relative Market Presence	185
10.3	GIS Data Conversion (Focus) Contractors	186
10.4	Aerial Mapping Firms	186
10.5	Engineering Firms	188

10.6	Automated Conversion-Oriented Contractors	189
10.7	GIS Vendors	190
10.8	Criteria for Selecting a GIS Conversion Contractor	191
	Technical Considerations	191
	Company History	191
	Full-Service Company	192
	Company Location	192
	Organizational Structure	192
	Corporate Ownership	192
	Company Revenue Base	193
	Company GIS Commitment	193
	Company Resources	193
	Personnel Experience	193
	Company Experience	194
	Technical Plan of Operation	194
	Price	196
	Data Conversion Cost Factors	196
	Number of Contractors	197
	Target GIS	197
	Data Quality	197
	Database Design	198
	Source Document Scrub	198
	Source Documents	198
	Conversion Deliverable Products	198
	Representation	199
	Labor Costs	199
	Quality Assurance	199
	Postconversion Processing	200
	Project Schedule	200
	Competitive Bids	200
	General Bidding Aspects	201
	Database Complexity	201
	Potential for Other Services	201
	Proximity to the Client	201
	Schedule Flexibility	201
	Shifts	202

10.9	Risk	202
	Financial Report	202
	Lack of Experience	202
	Loss of Key Personnel	202
	Data Quality Problems	203
	Data Translation Experience	203
	Poor Project Management	203
	Inadequate Data Conversion Software	203
	Inadequate Workstation Capacity	204
	Failure to Meet Schedule	204
	Staff Acquisition Problems	204
	Market Segment Experience	205
	Pending Litigation	205
10.10	General	205
10.11	GIS Data Conversion Trends	206

11 Impact of Emerging Technologies — 209

11.1	Introduction	209
11.2	Data-Driven Approaches	211
11.3	Scanning and Related Technologies	213
	Automated Vectorization	213
	Interactive Vectorization	215
	Raster Data Utilization	216
	Raster Land Base	216
	Reference/Detail Drawings	217
	Raster Maps	217
	Incremental Map Conversion	217
	Hybrid Files	217
11.4	Field Data Collection	218
11.5	Global Positioning Systems and Conversion	220

12	**Developing a GIS Data Conversion Plan**	**221**
12.1	Introduction	221
12.2	Identifying Sources of Information to Be Converted	222
12.3	GIS Database Design Factors	222
12.4	Other Factors	225
12.5	Identifying Existing Data Sources	225
12.6	Identifying New Data Sources	226
12.7	Conceptual Database Design	227
12.8	Physical Database Design	228
12.9	Internal vs. External Conversion	229
12.10	Purchasing Data Conversion Services	230
12.11	Source Preparation and Scrub	234
12.12	GIS Data Conversion Work Plan	235
12.13	In-House Activities	241
12.14	Quality Assurance	241
12.15	Change Control	242
12.16	GIS Data Maintenance	243
12.17	Pilot Project	244
	Glossary	247
	Index	283

Preface

Throughout the years, UGC Consulting has helped its clients understand one of the most critical and expensive aspects of geographic information system (GIS) implementation—data conversion. UGC Consulting concluded that the market would be well served with a noncommercial, comprehensive work that addresses the technical and management issues surrounding successful data conversion projects. With the publication of the *GIS Data Conversion Handbook*, information on this subject is now available in one comprehensive volume.

The authors believe that broader acceptance of GIS depends on the contributions of knowledgeable individuals to an ongoing exchange of information. From the earliest planning stages, this work has represented the collective knowledge of many of UGC Consulting's GIS conversion specialists and has been reviewed by industry peers with the highest level of expertise in this discipline. Some 40 experts contributed their ideas, observations, and technical experience in data conversion to writing and editing the 12 chapters of the *Handbook*. The goal of the *Handbook* is to help organizations assimilate this technology as easily as possible and to realize tangible benefits from their data conversion expenditures.

A special effort has been made to present a well-rounded and unbiased reference source on data conversion, without commercial endorsements of products, vendors, or conversion contractors. A major consideration has been to provide the *Handbook* with an extended shelf life so it is relevant for several years. This has not been an easy task because GIS is a rapidly growing and changing technology. The authors have strived to achieve currency by concentrating on data sources, markets, issues, and aspects of data conversion management relevant to the present and the future.

The *Handbook* begins with background information about the broader realm of GIS and an overview of public, regulated, and private market applications that employ GIS technology. The *Handbook* examines the hardware, data conversion sources, methods, and other subjects necessary for the reader to gain a conceptual understanding of GIS data conversion.

Each reader must realize that much of the information contained herein is presented in a generalized fashion. In the two or more decades that commercial GIS data conversion has existed, it has been proven that each project is unique. No two projects have the same set of intended applications, requirements, specifications, source documents, quality control standards, or business mission. However, this text investigates the key themes and lessons that can be applied to a broad range of projects while avoiding the suggestion that a common solution is universally applicable or available.

Each chapter of the *Handbook* builds on the information contained in the previous one. At every step the reader is introduced more and more to the complexities and interrelationships of data conversion. The last chapter, "Developing a GIS Data Conversion Plan," reviews the major components of a typical data conversion plan and puts them into perspective by providing a general overview of the conversion process and of the critical stages to be managed.

The authors believe the information in this book is invaluable in helping readers gain insight into the many pitfalls encountered in a data conversion project. In addition, the *Handbook* is a useful reference work for helping to educate executives and staff personnel to the complexities and risks of a GIS data conversion effort. We sincerely hope you find the *GIS Data Conversion Handbook* a useful addition to your project management library.

Glenn E. Montgomery
Harold C. Schuch
Denver, Colo., USA

Acknowledgments

This handbook is an outgrowth of the collaborative efforts based on work recently completed or underway on more than 100 complex GIS data conversion consulting assignments worldwide for utility, telephone, government, transportation, natural resources, and various commercial applications.

The principal contributors to this effort, apart from the authors, are many top consultants at UGC Consulting, including Craig Bachmann, Ron Cooper, Kevin Corlis, Tony DiMarco, Bart Elliott, Jim Hargis, Steve Hick, John Kelly, Brian Kiernan, Ed Odenwalder, Stan Weber, and Dean Zastava.

All-important prepublication reviews were provided by Ed Forrest, editor of the *A-E-C Automation Newsletter*; Dr. Michael Goodchild, Department of Geography, University of California, Santa Barbara; and Jerry Williams and Ed Sangaline of IBM. Their constructive technical reviews are fully acknowledged and greatly appreciated. The authors also wish to acknowledge several organizations for their contribution of photographs and figures that were used throughout the text, and additional manuscript reviews. These include Baymont Technologies, Cartotech, and MSE Corporation.

Last, but not least, the authors wish to thank Dr. H. Dennison Parker, founder of GIS World, Inc., Fort Collins, Colo., and key members of his staff, including Kimberley I. Parker, marketing and promotion; Darin E. Sanders, art director; Wade L. Smith and Darin E. Sanders, graphics and production; Christine Thompson, production manager; and Christine Thompson and Darin E. Sanders, cover design. We also acknowledge the contributions of Sharolyn Berry Eitenbichler for copy editing and Rebecca Herr and Jason Bovberg for proofreading. Dr. Parker's support in the publication of this book was essential and in the spirit of his commitment to the worldwide GIS community.

List of Figures

1.1   Major components of the data conversion process.
1.2   Format of a relational database table.
1.3   GIS data layers.
2.1   Sample project schedule.
2.2   Fundamentals of GIS data conversion.
2.3   GIS implementation costs.
2.4   Developing a data conversion approach.
2.5   Topdown database design.
2.6   Converting different data sources.
2.7   Source data matrix.
2.8   Workstations at a conversion contractor's facility.
2.9   GIS data conversion project organization.
3.1   Typical hardware configurations in the 1970s to early 1980s.
3.2   Typical hardware configurations in the late 1980s to 1990s.
4.1   Portion of a municipal planning map.
4.2   Portion of an electric utility map.
4.3   Portion of a poorly drawn telephone facility detail.
4.4   Portion of a wetlands map.
4.5   Portion of an aerial photograph.
4.6   Aerial map representing photograph in figure 4.5.
4.7   Sample water service card.
4.8   Portion of a facilities record.
4.9   Portion of a DLG file.
5.1   Two-panel land/facilities data.
6.1   Map digitizing.
6.2   Map digitizing workstation configuration.
6.3   Map setup flowchart.
6.4   Keyboard data entry.
6.5   Photogrammetry process.
6.6   Scanning operations.
6.7   Conventional field survey.
6.8   GPS satellite.
6.9   Field inventory operations.

7.1	Absolute accuracy.
8.1	Database design stages.
8.2	User needs survey.
8.3	GIS database elements.
8.4	GIS hierarchical data model.
8.5	GIS relational data model.
9.1	Typical multiparticipant organizational structure.
11.1	Process-oriented conversion strategy.
11.2	Open nonsystem-specific database utilization.
11.3	Geocoding of structures.
11.4	Generating graphics from nongraphic databases.
11.5	Theoretical automated scanning process.
11.6	Typical hybrid raster/vector solution.
12.1	GIS positional accuracy vs. cost.
12.2	GIS database purchase cycle.
12.3	Guide to data conversion.

List of Tables

1.1	Typical GIS land elements and attributes.
4.1	Map type.
4.2	Map content.
4.3	Map use.
4.4	Standard sheet size nomenclature.
5.1	Typical land base data.
5.2	Sample municipal facilities data.
5.3	Sample telephone facilities data.
5.4	Sample electric facilities data.
5.5	Sample gas facilities data.
6.1	Horizontal and vertical accuracy levels.
7.1	Map scale usage matrix.
7.2	Accuracy requirements matrix.
7.3	Relationship of compilation scale and absolute positional accuracy (based on National Map Accuracy Standards).
7.4	Typical GIS positional accuracy requirements.
7.5	Positional accuracy requirements vs. GIS data conversion costs.
8.1	Sample GIS symbology.

Chapter 1

AM/FM/GIS and Their Markets

1.1 Introduction

This publication is designed to introduce the reader to the complex process of building databases for geographic information systems (GIS). This process is called *data conversion,* and it generally consists of converting existing information, predominantly paper maps and records, to a digital format that can be used directly by the target GIS. Figure 1.1 provides a simplified view of what the term *data conversion* means.

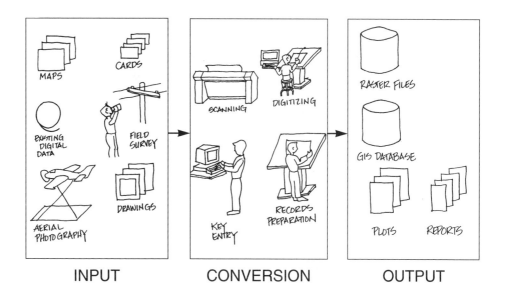

Fig. 1.1. Major components of the data conversion process.

The principal aspects of GIS data conversion and various data conversion related components of GIS are described in this publication. Each chapter presents a specific area of interest, and chapter 12, "Developing a GIS Data Conversion Plan," brings the information from all preceding chapters together. Specialized GIS terminology is presented and defined in a glossary.

1.2 Origins of GIS Technology

One of the interesting things about GIS is that it has such a varied background. In the late 1960s and early 1970s, several development efforts paralleled each other; the resulting products are now collectively known as GIS. Whether GIS development took place in the computer graphics lab at Harvard University; in a public utility company in Colorado; in a gas company in Houston; in the offices of a county in North Carolina; within natural resource management agencies in Canada; or as a spin-off of computer-aided drafting (CAD) or computer-aided drafting and design (CADD) technology, one thing is certain: private companies, organizations, and agencies looking for solutions to their business problems initiated these developments. This important underlying premise still exists today: GIS was started to meet the needs of organizations and agencies from several different market sectors, and especially to satisfy the need to handle and utilize geographically referenced data.

The birth of GIS is recognized as the successful completion of a research project at Harvard University in the late 1960s. This project resulted in the capability to shade slope maps with the help of an impact printer programmed to perform multiple striking. The program was called SYMAP, and it led to the formation of Harvard's computer graphics lab. In the 1970s, the lab produced a product called Odyssey that introduced a solution for polygon processing and polygon overlay operations. These products were the first to be identified as having GIS functionality.

Automated mapping and facilities management (AM/FM) appeared as a truly market-driven technology when Public Service Company of Colorado (PSCo) began conceptualizing an AM/FM project in the late 1960s. Public Service Company of Colorado then recognized the need to relate its computerized tabular data with a data layer containing the local geography and specialized utility network information. In the late 1970s, PSCo and IBM joined forces to build a prototype AM/FM system that became the Geofacilities Data Base Support (GDBS) system.

Generally, AM/FM was adopted by the utility industry to solve its growing mapping and facilities management needs. The first solutions were purely graphical and emphasized computerized drafting functions. During its infancy, AM/FM technology relied mostly on the representation of point and linear features such as poles and conductors, valves, and pipes. Related information, such as conductor voltages and pipe sizes, was stored as text in the graphic files without a database relationship established between, say, an electric conductor and a corresponding voltage. This early orientation toward utility applications caused a heavy emphasis on line-type data, and early AM/FM solutions were not well suited for the processing of area-type data.

In contrast, the first GIS solutions were oriented to applications involving areas within which some common trait(s) existed, such as soil types and/or tree stands. The beginning of GIS is generally recognized to have been at Harvard, when geographic information was used for planning and area analysis applications. Early GIS solutions were area-oriented and had very few line-based capabilities. Graphic editing functions were also weak or nonexistent.

As they evolved, AM/FM and GIS began to add competing functionality; however, both technologies experienced difficulty in shedding original limitations. Even today some AM/FM systems do not have elegant polygon solutions, such as polygon intersections or unions, and some GIS products have difficulties in providing efficient and effective graphic production functions such as digitizing, graphic editing, and plotting.

At present, vendors of GIS or AM/FM systems are attempting to diminish the appearance of any difference between GIS and AM/FM functionality. The system vendors are making an effort to provide all the major functions a client needs, regardless of whether the applications require AM/FM or GIS functionality. This has created the use of the term *AM/FM/GIS*. Currently, the marketplace appears to be replacing this term with GIS to mean all aspects of AM/FM/GIS. Throughout this publication, the term GIS is used to refer to all aspects of AM/FM and/or GIS.

Because GIS has graphic display and output functionality, and because the graphic database is much more powerful than conventional mapping systems, GIS has made inroads into the world of mapping to such an extent that many GIS applications have replaced the need to refer to conventional map formats and conventional map areas and scales. These newer systems offer *seamless* database capabilities for continuous graphic coverage of an area, without map boundaries or breaks. They are very efficient in the graphic production of maps; no longer must the user piece together separate data sets or be limited by map scales or the amount of information drafted on a map sheet.

Most organizations manually create and utilize a variety of maps in their daily activities. For example, public works and engineering departments use detailed street maps to show locations of streets, sewers, water mains, and other facilities. Thousands of drawings based on maps may be necessary to keep water distribution systems or utility transmission and distribution lines operating safely and efficiently. Planning departments utilize street maps, drawing zoning districts, and land use characteristics on them. Other street maps, some at different scales, are used to keep track of demographic information based on special geographic areas such as census tracts and electoral reapportionment districts. Police and fire departments utilize maps to delineate precincts and to determine optimal dispatch routes. Telephone companies use maps to resolve client assignment record conflicts within large wire centers.

In addition to maps, organizations utilize large quantities of location-specific information that can be related to a map. This type of information usually comes in the form of a nongraphic record or a graphical sketch. Examples include property ownership and taxation records, sewer or other utility maintenance information, equipment records such as manhole and pole cards, and census data on population within census tracts and blocks.

The quantity of information that needs to be created and accessed is overwhelming: some utilities and local governments have amassed several million pieces of geographically referenced paper, all of which have to be stored, maintained, retrieved, and used by many different people in different departments and locations. Distributing this information, often the original records, increases the probability that the information will be destroyed or lost. This is a major problem with manual records; often the solution implemented is restriction of access to these records. This solution usually has negative implications since limiting access to records erodes their usefulness.

Since the 1970s a variety of organizations have moved away from paper-based operations and have begun placing geographic-related, record-type information into tabular digital databases. But only recently have organizations and agencies begun to modify these tabular databases to add geographic-related match keys or database links that allow application software to relate information in a tabular database to information in the graphic database and vice versa. This functionality is usually based on a combination of selected CAD functions inherent in the graphic database, tabular database access functions, and the addition of special GIS applications. As a result, it has become increasingly practical and economically feasible to utilize GIS technology to help create, maintain, manage, and use maps and geographic-related records simultaneously.

Computers are used to automate drafting.

To further familiarize the reader with the differences among these basic technologies, we provide the following descriptions for the frequently used acronyms CAD, CADD, DBMS, AM/FM, and GIS.

Computer-Aided Drafting and Design (CADD or CAD)

The first commercially feasible applications based on graphic databases were developed to support a variety of design and drafting tasks, such as designing and drawing mechanical parts or architectural building plans. This led to the general acceptance of computer-aided drafting (CAD) and computer-aided drafting and design (CADD) systems, which utilize computer graphics technology to automate the traditional drafting and design techniques required in engineering and technical drawing. Recently, the term *CAD* has begun to refer to both CAD and CADD; therefore, CAD is used throughout this book.

Computer-aided drafting systems usually allow for the drafting of various drawing layers that can be selectively displayed and edited. Drawings using CAD can be three-dimensional models or simply a digital replication of manually drafted drawings. The simple digital replications do not allow for manipulation of textual or tabular information, except as graphic elements for display purposes only.

Extensive preparations for data capture are not required by CAD systems, a major difference from GIS. When using CAD, the user can begin creating a digital map without first having to complete and code an extensive database design for capturing the map. This is one reason why CAD systems can be used to create graphic data for a GIS, even if the data will have to be translated into the GIS format in a subsequent process. The strength of the graphic editing functionality of the CAD system can make the conversion of map data easier and more convenient, especially if the GIS does not have robust graphic editing functions.

Because of ease of use and similarity in function to manual design and drafting techniques, CAD systems have enjoyed a great popularity during the past 15 years. Virtually every architectural, engineering, manufacturing, surveying, and civil engineering firm in the United States uses some form of CAD. Many of these firms' clients now require drawings, plans, maps, and other drafted documents to be produced and delivered in CAD formats because design changes can easily be made and new drawings can quickly be produced.

Database Management Systems (DBMS)

Database management systems (DBMS) allow for the manipulation of digital tabular data. For example, the databases maintained by banks containing names, addresses, and account data are tabular databases.

A tabular database and access to it are controlled through software and a database design that collectively are called a *database management system* (DBMS). This means that the database contains user data, as well as a database design (schema) that the user has to create. The schema contains information regarding the number and type of data that the database holds. This includes the names of data fields (attributes), such as "phone number" or "address"; the data format of the attributes, such as "real," "integer," or "date"; the relationships/linkages between data, such as relating a client name with a phone number; and the grouping of data, such as placing all information for a single client into one record, "client account." The user creates the schema through a process called *database design*. The database design process must be completed before any data can be placed into the database and before the DBMS can function as expected to support an organization's or agency's applications.

The first solutions that were offered in DBMS were based on connecting one piece of information to another in a treelike structure.

These DBMS were called *hierarchical* and this solution was fairly successful in a broad range of data processing applications, including some GIS and AM/FM implementations. However, it was difficult to relate information from one branch of the data hierarchy to another branch. In other words, the data access was limited, and users could not extract related information easily unless it was stored together as a result of the database design.

The dynamic relationship limitation of hierarchical DBMS was overcome with the development of *relational* DBMS, or RDBMS. This solution is based on the creation of tables of data that hold related information in columns. For example, a table called "client account" would typically have a first column called "client name," a second called "account number," a third called "address," and so on. Each column is an attribute type, and each row in the table is an occurrence or record (i.e., a specific client). Relational DBMS are very flexible and typically allow tables to be dynamically joined to create new relations that a user requires to support new applications. Figure 1.2 depicts a sample RDBMS table.

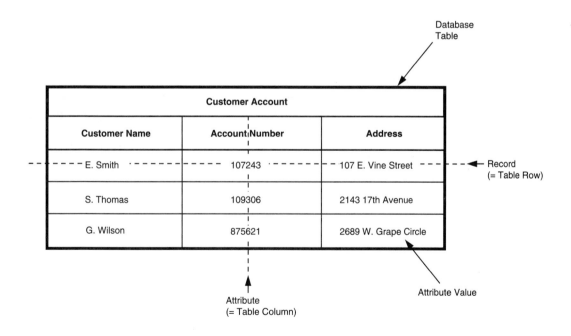

Fig. 1.2. Format of a relational database table.

The strength of the tabular approach lies in the functionality of the RDBMS, which allows the temporary or permanent linking of two or more tables. Therefore, it is possible to access and join information temporarily during its use, allowing the creation of data relationships that were not considered in the original database design. Nevertheless, RDBMS also require good database designs to ensure efficient processing and fast access to frequently used data. Databases must be carefully designed so that they can hold specific information and so that meaningful information can be extracted from them. A DBMS or RDBMS allows for the selection of portions of an entire data set based on certain selection criteria (e.g., list all homeowners with housing values greater than $60,000).

The most popular DBMS in use today are RDBMS because of this database access flexibility. Organizations and agencies implementing GIS for the first time and those migrating to their second generation of GIS are typically selecting RDBMS. The major drawbacks associated with relational database architectures, storage space requirements, and lengthy display times have been substantially mitigated by technological advances.

However, hierarchical structures are still found in modern relational database designs. If the user wants to create an "object" in the database such as a "hydrant" (represented by a symbol; has a geographic location; carries a set of attributes; has a certain color, rotation, manufacturer, and size), then these relationships are superimposed onto a relational database based on a hierarchical structure. This superimposition is done by an additional software module which is, in essence, another type of database manager. The need to control the functioning of a DBMS or an RDBMS to gain additional GIS functionality forced GIS vendors to develop additional software modules to control database managers. These software modules have not been generically named yet; most GIS vendors have introduced their own proprietary software module names.

The use of database management systems by organizations and agencies has increased dramatically in recent years. Many DBMS applications have been developed to manage information including geographic records; however, DBMS technology is still generally thought of as a tool to manage tabular data. Databases used by organizations and agencies in the future will most likely have a geographic component or layer.

Automated Mapping (AM)

Similar to CAD, automated mapping technology utilizes computer graphics technology to produce maps. Automated mapping often uses

CAD-type functions for digitizing maps (i.e., converting a manual map to digital form) and for editing, but in contrast to CAD, automated mapping usually does not offer a comprehensive set of engineering design functions to the user. Automated mapping does offer very strong graphic capabilities for the preparation of high-quality maps, even cartographic-quality maps. Automated-mapping technology also generally utilizes limited DBMS technology to relate tabular data to a mapped area for the production of statistical or choropleth maps. Because AM systems keep maps and associated information in digital form, all the cataloging, indexing, and organizational capabilities of electronic automation and software can be used to manage the production and maintenance of the maps and data.

Automated mapping was quickly accepted by a variety of organizations for several important reasons. One reason is that a central graphic database can easily produce maps of varying content, format, and scales as necessary for all the users. Problems caused by an individual walking away with, losing, or misplacing an original map (which happens often in a paper-based environment) are eliminated. Also, with an AM system, digital maps can be more easily updated as changes occur, thereby eliminating the lag caused by waiting until enough changes have accumulated to justify redrafting. As a result of AM, map maintenance can be less expensive and performed in a more timely manner.

Facilities Management (FM)

Many organizations own and operate geographically dispersed facilities, such as electric lines, gas lines, sewer lines, telephone cables, traffic lights, and/or water mains. These facilities must be efficiently planned, designed, constructed, operated, maintained, and managed. A water utility has thousands of pipes, each having a specific function, size, material, location, installer, maintenance record, etc. A city's traffic control system must keep track of hundreds of traffic lights, their individual location, timing sequences, etc. A city's public works department must keep track of its maps and geographic records to effectively construct and manage the city's diverse infrastructure.

Specialized automated systems have been developed that can be used to help an organization manage its facilities. These facilities management (FM) systems commonly utilize DBMS technology to allow the manipulation and query of information on facility installation, maintenance, type, capacity, manufacturer/supplier, and so on. Most of these systems were implemented by organizations during the 1960s, 1970s, and 1980s, without having related graphic database information stored in digital map form.

The FM systems were created where cards and records were entered into a database. Occasionally, this information was cross-referenced with paper maps via the use of match keys. A match key is a unique piece of datum that is common to the FM database and to the paper or digital map. Examples of match keys include structure numbers (pole number and manhole number), street addresses, and equipment or device numbers (valve number and switch number). Several specialized applications have been developed to run against FM databases, such as work estimating, bill of materials generation, and infrastructure audits.

Automated Mapping/Facilities Management (AM/FM)

Automated mapping/facilities management systems utilize a combination of automated mapping functionality and facilities management database functionality to create, store, retrieve, manipulate, and display a variety of land base and facilities information. These systems combine facility location information (AM) with facility record information (FM) as well as provide general land base information to which the facilities are referenced. The CAD concept of layered information is utilized in an AM/FM system, where each layer represents a different set of map information. For example, one layer may show the street network; another, city boundaries; and another, the sewer network. Also, spatial relationships for connectivity are stored to analyze flows of water, traffic, electricity, and/or gas between points in a network or to analyze overall network conditions.

Automated mapping/facilities management systems have strong linear/network graphic capabilities, and since they were originally developed by CAD vendors, a variety of engineering and design software modules are available to enhance them. Recently, most AM/FM systems have been enhanced with additional capabilities to support polygon processing, the ability to deal with and analyze areas such as parcels of land, planning zones, environmentally sensitive areas, and soil-type polygons. These enhancements have brought the functionality of AM/FM considerably closer to that of GIS.

The FM component of an AM/FM system means that database design is an important and necessary step before an AM/FM system can be constructed and used. The database has to be designed to model the real-world environment it is intended to represent, and be structured for the type of data it is intended to hold and the type of applications it is intended to support. This database design requirement makes an AM/FM system more complex and more difficult to implement than a CAD or an AM system.

Geographic Information Systems (GIS)

Like an AM/FM system, a geographic information system (GIS) utilizes automated mapping and DBMS or RDBMS technology to relate data to digital maps and to allow for the creation, storage, maintenance, retrieval, analysis, and display of various geographic and tabular information. As in an AM/FM system, layers of map information can be stored and accessed for analysis and display. A GIS has the capability to use any logical combination of data layers for analysis. Representative sample layers are shown in figure 1.3, which illustrates how several types of information can be separately stored, positioned over a common geographic area, and used together in analyses.

At the most fundamental level, an AM/FM system typically views a polygon (e.g., a parcel) as being an area enclosed by *n* lines. A GIS views a parcel as a single area constructed by edges having that area identifier as an attribute. The edges in such a topologic structure are constructed from points.

A GIS is similar to an AM/FM system, but because of its *area analysis* origin, it has strong and specialized polygon processing capabilities that support automated analyses between separate map layers. The spatial search functions of GIS include creating *buffers* around properties or other features, analyzing characteristics within a specified radius of a point, and determining the proximity of one feature to another. Overlay functions of GIS include calculating the areas that result from combining map features and/or layers (e.g., acreage of a tract of land having a specific land use and a certain zoning type) and displaying, in different colors, facilities that cross specific soil types.

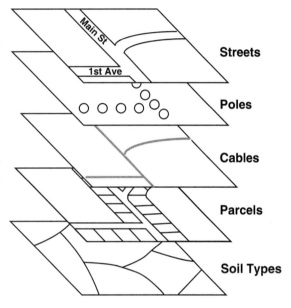

Fig. 1.3. GIS data layers.

In the world of AM/FM, standardized map products will most likely be produced on a routine basis. A GIS is characterized by the frequent use of ad hoc queries and unique, one-of-a-kind output products. For example, a local government might produce a specialized map showing several potential sites for a proposed landfill operation. To do so would require a combination of queries and spatial analysis steps. A utility using an AM/FM system is more likely to frequently produce a standard distribution map, for example, in its day-to-day engineering and operations groups. Although the definitions of *AM/FM* and *GIS* continue to be the source of some controversy, the utility sector tends to view itself as an AM/FM-oriented group, while local governments and groups involved in the management of natural resources tend to think of themselves as GIS users.

Typical GIS functions require spatial relationships to be stored among specific map features during the database design. These spatial relationships are similar to the connectivity of linear features required in AM/FM systems, but are extended to include features representing areas. These relationships are known as topological relationships, or collectively as topology. The implementation of *topology* or other information linkages in a database is called adding *intelligence* to the database.

As with the AM/FM system, database design is an important and necessary step that must take place before any data can be placed into a GIS and before the GIS can be used. The need to define the database structure so that topological area relationships are correctly modeled suggests that the setup of a GIS could be more difficult than the setup of an AM/FM system; however, since both types of systems are beginning to provide similar capabilities, the differences are quickly diminishing. In addition, some system vendors are providing menu-driven database design modules that greatly reduce the effort needed to perform this task.

1.3 Potential Benefits of a GIS

Two types of benefits can be realized with the use of a GIS. Quantitative or *hard* benefits are those benefits that can be measured and quantified in specific economic terms. Qualitative or *soft* benefits cannot be directly quantified or measured, but can ultimately impact the economics of a GIS project. The following are potential benefits that can be expected upon implementing a GIS.

Quantitative Benefits

Quantitative benefits differ considerably from one organization to another because of the variety of business functions performed, the current level of information automation, and the number and complexity of records/information problems that exist. The identification of quantitative benefits is usually based on detailed studies that identify the types of data the organization uses, how these data are maintained, the quality of the data, and the quantity of data. Some high-level quantifiable benefits are general enough that most organizations will realize them. Such benefits include the following:

1. Measurable increases in productivity during the creation, maintenance, and seeking/verifying of geographic-related information. A GIS automates routine, repetitive tasks, leaving more time for staff to focus on analysis and problem solving.

2. Support of real-time maintenance of geographic-related data and the distribution of updates to all relevant groups within an organization in a timely manner. This improved access to current data increases the overall effectiveness of information users.

3. Increased speed allowing very large volumes of data to be manipulated and analyzed. Faster analysis allows for additional alternative solutions to be considered.

4. Centralized database to provide a single source of common land and facilities information. Centralization will enable faster retrieval and selective modification of information and provide more consistent operations, including standardization, since all users will have access to the same current data.

5. Availability of faster technology that can eliminate map and record posting backlogs and accelerate work processes (e.g., processing of permits).

6. Ability to easily produce reports, statistics, etc., based on user-specified parameters to comply with legislation (e.g., reassessment and Environmental Protection Agency requirements) or public utility board requests for information.

7. Improved ability to exchange geographically oriented data with other entities or governmental bodies.

8. More efficient asset and inventory management.

9. Assistance in attracting new clients by locating desirable sites in terms of water availability, sewer service, industrial sites, advantageous labor pools, proximity to transportation, and other factors.

10. Support of engineering, planning, and real estate functions using current planimetric, topographic, hydrographic, land use, and tax map information.

Qualitative Benefits

Qualitative benefits cannot be used to directly justify a GIS acquisition from a cost point of view, but they can be used to strengthen the quantitative case. Again, qualitative benefits differ from one organization to another, and they typically have to be identified in detail through extensive analysis. Qualitative benefits for GIS generally include the following:

1. Improved responsiveness to inquiries through increased information accuracy, improved response time, and ability to quickly analyze larger volumes of data (e.g., complaint tracking).
2. Improved overall quality of analysis based on a continuous, accurate, and current digital database.
3. Uniformity of standard map products throughout an organization.
4. Capability to produce specialty maps at any desired scale.
5. Reduced information storage requirements (in a variety of physical forms).
6. Ability to evaluate a greater variety of design or planning alternatives.
7. More effective analysis of geographic-related data using techniques either not practiced or not available in a manual environment.
8. Improved economic development activities for municipalities and regional governments by allowing ready access to a variety of high-quality decision support data.
9. Faster data analysis tools and specialized products that greatly enhance and expedite management's decision-making capabilities.
10. Ability to identify and track changes in political boundaries, population centers, and other areas.
11. Improved emergency preparedness.
12. Assisted planning of optimal routes or rights-of-way for lethal commodity utilities (e.g., electric transmission lines).

1.4 Typical Database Information Layers

A GIS frequently contains many different types of information within a common database, sometimes as layers (as shown in fig. 1.2), and other times as a combined database where data can be separated during display or analysis. Apart from the design needed to administer the information within a database, GIS information is comprised of graphic representations (i.e., digital map features) and associated tabular record information (i.e., attributes). Table 1.1 contains a high-level list of typical land base features and associated attribute data.

While no single database necessarily contains all the data elements presented here, a typical municipal database will contain most of

Table 1.1. Typical GIS land elements and attributes.

Element	Attribute
Street Centerline	Name Classification Pavement Type Bus Route Snow Removal Sand Route Speed Limit Low-Left Address Low-Right Address High-Left Address High-Right Address
Road Right-of-Way	
Alley	
Bikeway	
Pathway	
Fence	
Tree	Type Historic
Bridge/Overpass	
Trail	Type
Pavement	Type Resurfaced Condition
Edge of Pavement/Curb Cut	
Driveway	
Railroad Track	Owner
Parking Lot	Type Owner
Parking Garage	Type Owner
Transportation Center	Type Owner

Table 1.1. (continued)

Element	Attribute
Building	Type Stories Use Vacancy No. Apartments Substandard Owner Tenant Historic Address Building Permit Fireplace Permit Septic Tank Permit
Condominium	Type Use Vacancy Substandard Owner Tenant Address Building Permit Fireplace Permit
Pool	
Park	Name Type Jurisdiction
Landfill Site	Name Type Jurisdiction Owner Operator Status
Hazardous Waste Site	Name Type Owner Operator Status

these, and many more. An electric utility company will add elements such as poles, cables, transformers, switches, and fuses. A gas utility company will add gas mains, regulators, valves, and cathodic protection devices. A public works department may desire to work with topographic data such as spot elevations and contour lines, which require that the GIS database be able to handle three-dimensional data.

Each GIS database is unique and is designed and constructed to meet the complex information management needs of the specific organization it serves. These differences necessitate a careful database needs assessment prior to each data conversion project.

1.5 GIS Market Sectors

What types of organizations have implemented or might implement GIS? What market segments do they represent?

There are a variety of user groups within several distinct market sectors that need to buy or build data for use in their GIS. The GIS market can be separated into three primary market sections. Each group has its own requirements, such as accuracy, map scales, general applications, data types, GIS procurement methods, and so on.

1. **Public Sector**. This group includes federal, state, and local government agencies.

    -Federal Government
      Agencies
      Defense
    -State Government
      Public Works/Transportation
      Natural Resources
      Environmental Management
      Highways/Transportation
      Lands/State Planning
    -Local Government
      Tax Assessment
      Planning
      Management Information Systems
      EmergencyManagement Systems/Police/Fire

2. **Regulated Sector**. This group of buyers is made up of regulated industries and franchised utilities.

    -Electric   -Gas
    -Telephone   -Water
    -Sewer   -Cable
    -Storm

3. **Private Sector**. This group includes the following industries and organizations.
    - Timber and Forest Products Companies
    - Real Estate Companies
    - Retail Companies
    - Financial and Insurance Companies
    - Transportation Companies
    - Other Emerging Markets

1.6 Public Sector

The public sector user group consists of federal, state, and local government agencies. Each of these agencies can be further divided into specific subgroups that have common requirements and components.

Federal Government

Many agencies within the federal government utilize GIS data in their daily activities or in support of their public charter. The GIS data needs of these agencies are quite varied. From the U.S. Department of Defense, to the U.S. Department of Agriculture, to the U.S. Bureau of Census (USBC), the requirements for GIS data are wide ranging. Each agency has a specific mission that supports the goals and directions of the American people through the president and Congress.

As a purchaser of GIS data, the federal government primarily contracts for conversion services and conversion support services (e.g., photogrammetry). These contracts usually involve photogrammetric mapping, conversion of existing maps and records, and database development. However, the federal government also has considerable mapping and conversion capability of its own. The Defense Mapping Agency (DMA) is the world's largest map producer, and both the U.S. Geological Survey (USGS) and the U.S. Forest Service (USFS) have significant mapping capabilities.

Land positional accuracy requirements for federal agency GIS projects are usually low, with ±40 feet to greater than ±100 feet being satisfactory for most applications. Although the defense department's typical land positional accuracy is between ±50 to ±100 feet at map scales ranging from 1:20,000 to 1:250,000, some requirements need more accurate data, for example, the mapping of military bases and facilities. Land positional accuracy requirements for these applications are generally ±10 feet at map scales of 1"=50' to 1"=100' (1:600 to 1:1,200).

Perhaps the best-known maps from the federal government are 7.5 minute quadrangle maps (quads) produced by the USGS. Quad map scale is 1:24,000 (1"=2,000') with a land positional accuracy of ±40 feet. These USGS quads are a widely used data source for GIS applications with low land positional accuracy requirements. Since many quads are 10 years or more out of date, a requirement for current information must also be considered before deciding if these maps are a suitable GIS data source.

Federal land management applications for GIS include land use and vegetation mapping; minerals management; and mapping of topographic features, geologic features, timber stands, wetlands, soils, and flood zones. Other applications include analysis of census data, environmental assessment, wildlife monitoring, hazardous waste management, military base mapping, and facilities management. More than 100 separate GIS programs are under way within the federal government.

State Government

State government agencies also have many different requirements and needs for GIS data. The missions of these agencies are quite varied, ranging from natural resources management and environmental regulation to law enforcement, transportation planning and management, and commerce.

The implementation of GIS technology is a growing trend within many state government departments. In some cases different departments use identical GIS hardware and software, but often entirely different systems are implemented. This phenomenon creates different needs and requirements for GIS data.

Direct purchases of GIS databases by state government agencies are somewhat limited. Many agencies acquire data through agreements with federal government agencies or through service support contracts. The procurement process within these agencies is such that data purchases must be planned far enough in advance to allow requests for additional funds to be included in the budget planning process. As more and more state agencies utilize GIS technologies, the need for GIS data will grow. The principal sources of this data will include existing commercial databases whenever possible, and when not possible, conversion contracts.

Land positional accuracy requirements for state government agencies are usually broad because the applications are oriented to the entire state or to large geographic areas within the state. Often, land positional accuracies of ±50 feet to greater than ±100 feet are satisfactory for map scales between 1:24,000 to 1:50,000. State agencies frequently obtain data from the USGS and the USBC, who compile data with ±50 to ±100 feet land positional accuracies. In some cases, agencies within state governments have requirements for higher land positional accuracies. Departments of highways and transportation often need land positional accuracies of ±2 to ±10 feet for detailed profile designs and construction drawings that are created using internal resources. Departments of parks and recreation may also require fair accuracy data due to the site-specific planning, design, and management of park facilities for which they are responsible.

Because of the broad range of missions of departments in state governments, applications are varied. Typical applications include wildlife and forestry management, surveying and mapping of state lands, impact analysis of mining operations, and geological research. In some cases, coastal zone management and marine resources management are also required. Environmental regulation, hazardous waste management, road and highway planning and maintenance, and land use management and planning are other typical applications required by these agencies.

Local Government

With over 11,000 cities and counties in North America and the rapid growth in GIS technology implementation, there is an increasing need for GIS data. Prior to 1990, almost one-third of the cities and counties in the United States with a population of more than 75,000 had acquired a GIS. Many local governments with populations of less than 50,000 are looking at GIS technology, but have not yet implemented it because of the high cost (or perceived high cost).

Many different departments in local governments have a wide range of GIS data needs. Data requirements typically include information to support tax assessments, local planning and zoning, public works and engineering applications, emergency services, and environmental management.

The acquisition of GIS data usually involves hiring a conversion contractor to develop a current land base from newly acquired aerial photography. The contractor then performs the conversion of existing property maps and engineering drawings and fits this information to the land base. In some departments, commercial databases and/or USBC data are purchased. Census data are generally used to develop voter redistricting plans.

Positional accuracy requirements for local government departments are usually high. Land positional accuracies of ±5 or ±10 feet are common. Within public works departments, where engineering applications are common, the accuracies used approach the ±1 or ±2 feet level for certain civil engineering functions. Map scales for local governments are typically 1"=100' (1:1,200) but can range from detailed 1"=50' (1:600) maps up to more general 1"=2,000' (1:24,000) maps.

Typical local government applications include property tax assessment, permit processing, code enforcement, parcel mapping, land use planning and zoning analysis, facilities inventory and maintenance, infrastructure management, and emergency vehicle routing.

1.7 Regulated Sector

The regulated sector of GIS data buyers includes electric, gas, and telephone utility companies.

Electric Utilities

Electric utility companies use GIS information for planning, designing, constructing, and maintaining electric facilities.

Electric utilities were among the first users of GIS technology. They used GIS to automate the storage and retrieval of detailed maps and records associated with transmission and distribution facilities. Many electric utilities that installed a GIS in the late 1970s or during the 1980s are now updating their GIS to take advantage of new and improved technologies.

The positional accuracy requirements of this industry are broad. Generally, data accuracies of ±10 to ±50 feet are needed. Map scales for the electric utility industry are commonly 1"=100' (1:1,200) and 1"=200' (1:2,400), but range from 1"=50' (1:600) to 1"=2,000' (1:24,000). Transmission departments usually use smaller scale (larger area, less detail) maps, while distribution departments usually require larger scale (smaller area, more detail) maps.

Electric utilities often purchase off-the-shelf land bases such as Digital Line Graph (DLG), Topological Integrated Geographic Encoding and Referencing (TIGER), or existing GIS land base data to use as a reference layer for the placement of their facilities data. Existing maps can only be used as a source for GIS land base data when the utilities' GIS users have a high level of confidence in this information. Electric utilities also work with local government agencies to develop jointly funded GIS land bases.

Typical electric utility applications include system planning, load flow analysis, transmission line routing, vegetation management, map maintenance, client account analysis, work order preparation, underground facilities location, regulatory commission reporting, facilities inventory, and facilities management.

Gas Utilities

The GIS data needs of gas utility companies relate to the location and maintenance of gas transmission and distribution facilities. Gas utilities generally require data with high positional accuracy. Gas utilities are using GIS technology as a tool for network analysis, pipe repair/replacement, decision support, safety (leak survey), and regulatory reporting.

Gas utilities require detailed information about their transmission and distribution facilities, most of which are underground. Land positional accuracy requirements are usually in the range of ±2 feet to ±10 feet. In addition, accurate supplemental dimensions are also necessary to locate valves and fittings on maps. Stringent accuracy requirements are needed because these firms are responsible for managing a lethal commodity.

Map scales from 1"=50' (1:600) to 1"=2,000' (1:24,000) are commonly used by gas utilities. The smaller scale maps (1"=2,000') are typically used for rural areas and transmission line routes.

Gas utilities usually purchase conversion services for the development of their GIS databases. Where mapping needs involve transmission line routing over large areas, commercial land base data or USGS quads are usually purchased as a source for the land base layer.

Telephone Companies

Telephone companies require GIS data covering large geographic areas to assist them in managing their facilities and wire centers. Multiple map scales are utilized to map the service areas economically.

Land positional accuracy requirements are usually low, with ±50 feet meeting most needs; however, for some applications, accuracies in the ±10 feet range are necessary. Telephone companies are also beginning to show interest in multiparticipant projects that allow them to obtain higher land positional accuracy data at more reasonable costs. Given a more widespread availability of commercial GIS databases and U.S. government data, telephone companies are realizing that adequate GIS land bases can be developed at a reasonable cost.

General GIS applications for telephone companies include planning, design, construction, and maintenance. The ability to trace a wire pair throughout a wire center is one of the desired applications. Another important application is the ability to automatically update a complete database when a wire pair is reallocated (called *cable ripple*). In addition, inventory management, transmission analysis, and work order management applications are also important.

1.8 Private Sector

Private sector companies consist of a variety of commercial organizations, such as timber and paper product companies, financial services and insurance companies, real estate companies, transportation companies, retail companies, and other commercial users of GIS.

Natural Resource Companies

Mining Companies. Considering that mining operations can be based on vastly different methods of defining a single set of GIS-oriented requirements is difficult. But they all have a common element: the need for a three-dimensional database. Both underground and surface operations must describe underground ore bodies. Surface operations must calculate overburden removal and seam depth, and underground operations must be able to position extraction operations in three dimensions. Borehole information has a horizontal position and several depths for horizons, layers, seams, and ore body thickness; each of these is usually accompanied by more information, such as materials, concentration values, ore quality, and other data.

Therefore, mining operations have very special database needs, and this makes most commercial GIS difficult to use. The exceptions are those applications where two-dimensional solutions are sufficient. These cases are related to land/lease ownership and reclamation projects, such as when a company owns a series of leases that overlap ore bodies or when the company is required to perform a specific amount of reclamation work after operations have ceased at a certain location. Keeping track of the different royalty payments is much easier if a GIS is used to establish exactly who is affected by how much extraction. Small system houses handle other, more specialized applications such as processing drillhole data, geophone profiles, and ore body concentration profiles. Mining companies also use photographic and remote sensing imagery extensively, both on a photographic base and in raster form. This imagery may be collected from aircraft or satellites.

Oil Companies. For oil exploration/exploitation activities, companies require large databases containing many attributes at many discrete points (well locations). Database designs must accommodate the fact that not all locations have all the attributes present. Statistical analysis of data availability is required. Subsurface modeling from *top of horizon* data is required for display analysis and input to computation of *isopach* (seam thickness map) models.

Surface information, including well locations and oil lease information, must be mapped. Lease information includes property boundaries, owner name, lease name, inception and expiration dates, and so on. Well locations are displayed by symbols indicating type and status. Leases are typically stick representations without intelligence. Intelligence is usually carried on a parcel locator point within the lease boundary.

Timber and Forest Products Companies

Timber and forest products companies are major users of GIS data. These companies' applications relate to timber stand locations and species content, and include attributes such as tree heights, timber volumes, and disease detection. Access roads, topography (i.e., slope), and hydrological features are also needed. These companies require three-dimensional databases for the purpose of analyzing slope conditions, runoff/erosion potential, sedimentation, and exploitation economics.

Land positional accuracy requirements are broad, generally in the range of ±100 feet. Aerial mapping of timber stands and data captured from USGS quads are the primary GIS data capture methods utilized.

Map scales can range from 1"=100' (1:1,200) for individual stands to 1"=4,170' (1:50,000) for large areas. The use of GIS technology in this industry is growing. These companies need GIS to help manage their resources better, to help maintain profits, and to help them comply with state and federal environmental regulations.

Financial Services and Insurance Companies

Banks, insurance, and other financial services companies require GIS data for marketing and client service applications. Banks and credit companies link income levels, demographics, and population characteristics to geographic locations in order to identify areas for branch offices, potential clients, and other information.

Insurance companies use GIS data to identify flood zones, crime areas, and emergency services locations. This information will be used increasingly in calculations to determine insurance premium rates.

Data requirements for these applications are usually satisfied by commercial GIS databases or data that can be obtained from the federal government, such as census and topographic data (assuming acceptable currentness and availability), and sometimes from the local government. Land positional accuracy requirements are low, with ±100 feet being adequate. In some cases, even lower accuracy information will suffice. Map scales of 1"=400' (1:4,800) to 1"=2,000' (1:24,000) or greater are used.

Real Estate Companies

Real estate companies have an obvious requirement for GIS data. Their involvement with GIS technology to date, however, has been limited. Many larger real estate companies are now beginning to look at GIS as an enhancement of their daily applications and client services.

Data requirements are not extensive for these firms. Basically, they require simple street maps, property boundaries, and land use and address information. Land positional accuracy is not a major consideration due to the need for only approximate graphic representations. Land positional accuracy requirements typically are greater than ±50 feet, and map scales tend to be 1"=2,000' (1:24,000) or greater.

Real estate companies are among the first organizations to realize the full potential of hybrid raster and vector GIS data on a broad basis. The concept of being able to locate an address on a GIS map and then to display an image of a photograph of the building on the property appeals to nearly all real estate professionals.

Transportation Companies

There are many companies that deal with the transportation of goods and passengers. For these companies GIS data requirements include information on streets, addresses, highways, bridges, railroad tracks and crossings, and tunnels for land transportation operations. Air freight and sea transport operations require extensive facilities information and data on air traffic zones, airport and port facilities, sea lanes, harbor configurations, and topography.

Land positional accuracy requirements vary, from low for work with large geographic areas such as cross-country and transoceanic transport, to as high as ±20 feet to greater than ±100 feet for vehicle routing and navigation within cities and congested areas. Since transportation companies typically cover large service areas, map scales of 1"=2,000' (1:24,000) and greater are normally used. The integration of GIS with real-time, satellite-based vehicle and vessel tracking capabilities will be the next major developmental area for transportation companies. Data conversion requirements may be impacted by these developments.

Retail Companies

Market research companies provide services to large retail companies to assist them in locating new stores or facilities and to evaluate existing store locations in terms of sales volume, local demographics, and traffic patterns. The GIS data required to support these services include financial, demographic, transportation, commercial, and environmental parameters.

Land positional accuracy requirements are low, typically greater than ±100 feet. Land base data is usually obtained from commercial GIS databases containing streets and addresses. Greater land positional accuracy data may be required if the company's applications involve site location and planning for new facilities. The analysis/evaluation applications require that all marketing information is current and accurate. Map scales for the retail industry are usually 1"=2,000' (1:24,000) and smaller.

The classification of market sectors used in this chapter has gained general acceptance in the marketplace and is found in a variety of publications. There are many other users of GIS technology, but the sectors listed here represent the principal purchasers of data conversion services at present.

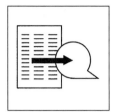

Chapter 2

GIS Data Conversion

2.1 Introduction

As illustrated in figure 2.1, the hardware and software components of a GIS can sometimes be bought and delivered in one month, but a functional populated database for most custom applications is typically not available until much later. Several years are often required to create the database for a large GIS project, and the creation can cost more than the GIS hardware/software itself. The creation of a GIS database, or GIS data conversion, almost always represents the most significant investment in any GIS project.

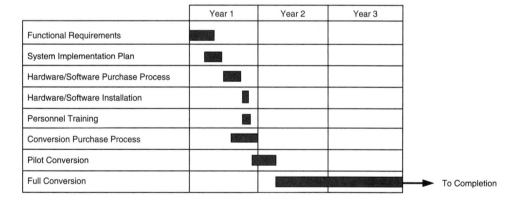

Fig. 2.1. Sample project schedule.

Data conversion involves populating a database to meet the requirements of a GIS. Typically, conversion transforms data from existing formats, such as paper records and maps, to the digital format required for a specific GIS product. Prior to data conversion, the database has to be structured to support planned application requirements and to model the desired environment. Entities, attributes, and data relationships must be defined through a database design process. Database design and the subsequent populating of the database with data represent the most important aspects of conversion.

Existing information to be used in the conversion process may reside on a variety of media. Paper records and documents, such as maps and index/record cards, represent common *sources*. Some existing information may also be stored as digital records in the form of CAD drawings or digital tabular files, such as assessor's records, which the user wishes to integrate with the GIS. Existing digital records typically need to be translated or reformatted as part of the data conversion process. Whether paper or digital, data is taken through a number of conversion processes before residing in the final GIS database.

Terms commonly associated with data conversion processes include scrubbing, scanning, vectorizing, image processing, digitizing, field inventory, stereodigitizing, control points, Global Positioning System (GPS), translation, validation, coordinate geometry (COGO), rubbersheeting, and warping. Please refer to the glossary for definitions of these terms.

2.2 GIS Data Conversion Issues

A number of issues must be addressed prior to embarking on a GIS data conversion project. Figure 2.2 illustrates the principal tasks and their relationships. These tasks are fundamental to any GIS data conversion project, and they comprise a baseline against which specific GIS data conversion questions may be asked.

As shown in figure 2.2, the data sources (maps and records) and the users are linked by tasks. Data sources are composed of land base (the map of the land in the area being converted) and facilities information (the cables, pipes, etc., that users overlay on the land base). The GIS operates between the user and the data sources. Converted data are loaded into the GIS. The GIS provides data access functions to the user. These functions create data requests to the GIS, resulting in information being returned to the users. In most cases, the link between the users and the data sources is replaced by direct interaction between GIS and user. In more than one sense, converting data sources into a GIS brings the data closer to users.

The mechanism used to make data requests to the GIS is part of the *applications*, which are program functions the GIS provides or has the capability to provide via user programming. The applications are

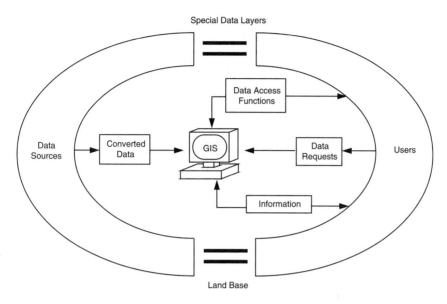

Fig. 2.2. Fundamentals of GIS data conversion.

developed as a result of a functional requirements study and subsequent requests from the users. The requirements study and requests, together with the database design necessary to run the applications, are called *requirements*.

For a GIS data conversion project to be successful, four fundamental issues have to be dealt with: the data requirements of the GIS, the data sources that are available, the techniques that are to be used in populating the GIS database, and how to "make it all fit together." These issues can be dealt with by answering corresponding sets of questions, as follows:

Data Requirements of a GIS

- What land base and facility data are needed to support the required and planned applications?
- What are the positional accuracy requirements for both the land base and facility data?
- What are the graphic and nongraphic database requirements?
- What are the required data relationships for a complete database design?
- What is the target GIS platform?
- What is the target GIS data format and database structure?

Sources of Data

- From what sources are the data to be obtained?
- Are the available data suitable for the GIS?
- Are data in a format that can be effectively converted to the GIS?
- How well will the facilities data fit the new GIS land base data?

Techniques of Populating a GIS Database

- Does the conversion approach meet the project requirements, and is it appropriate for the available source data?
- Is the conversion approach cost-effective?
- Should outside expertise and/or resources be used in the design and implementation stages?
- How will the converted data be maintained and by whom?

Making It All Fit Together

- Does the implementation plan meet the criteria for success?
- Does the project have executive support?
- Have risks been minimized?
- Have realistic and reasonable budget estimates been established?
- Is the schedule both appropriate and realistic?
- Have organizational roles and responsibilities been clearly defined?
- Is the organization prepared for the challenges and changes brought by a GIS?
- What checks and measurements are in place to assure all objectives are met?

Generally, data for GIS are a combination of land base features and attributes, appropriate facility features and attributes, and other types of data such as natural resources or demographic detail. Land base features are typically rights-of-way, edge of pavement, hydrography, parcels, building outlines, political or administrative boundaries, and other spatially oriented items of interest. Typical land base attributes include street names, addresses, pavement type, vehicle weight restrictions, traffic flow rates, etc. Facility features typically include utility network representations (e.g., poles, cables, pipes, valves, and transformers). Facility attributes may include size, type, date installed, manufacturer, and so on. Chapter 5, "Land Base Data vs. Facilities Data," provides more information on these types of data.

All fundamental data conversion issues must be addressed in sufficient detail to ensure a successful GIS data conversion. An organization may find it desirable to obtain external expertise to address these issues, as it is rare for internal staff to have the necessary skill levels at the outset of a GIS data conversion project. Organizations often augment internal knowledge of maps, records, and operations with external knowledge from consultants and conversion contractors. These external sources either assist in the creation of or actually provide data requirements, database design, database population, and applications implementation. If done successfully, conversion is a one-time operation. Most organizations cannot afford to redo a conversion because of the costs involved.

GIS Data Conversion Effort

The level of effort necessary for the data conversion portion of a GIS project can be determined by comparing data available from existing source materials to data requirements of the GIS. Data conversion is usually both lengthy and expensive, and has traditionally been the most expensive part of implementing a GIS. Figure 2.3 shows the approximate cost of conversion relative to the cost of other GIS components.

Estimating the total cost of a GIS data conversion effort is difficult due to the varied and complex nature of GIS. By analyzing the data requirements, sources, and available conversion techniques, an approximate cost can be derived based on comparisons to historical precedents. Generally, high-level estimates of conversion costs are calculated prior to the start of implementation. However, estimates must be refined to effectively plan and execute the data conversion process. In actuality, the most accurate estimates for a project come from the bid sheets included with the GIS contractors' conversion proposals; but by then, the decision to proceed with data conversion has usually been made.

As depicted in figure 2.3, the cost of conversion typically exceeds 50 percent of the cost of implementing a GIS. This is primarily due to the significant amount of conversion tasks that must be performed manually. While some automated data conversion techniques exist or are under development, requirements for GIS data conversion projects and the type and quality of source documents are so varied that it is impossible to design universally applicable automated conversion techniques.

Initiatives to reduce the traditional cost of data conversion have resulted in conversion contractors looking to technology for solutions. A number of automated techniques for populating a GIS are in use today, and several research and development projects are under way to

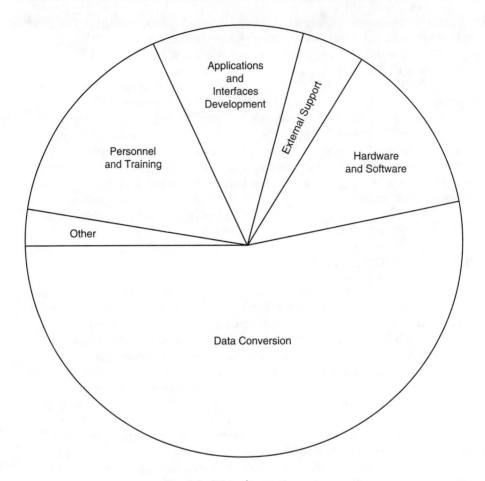

Fig. 2.3. GIS implementation costs.

develop cost-effective data conversion solutions. See chapter 6, "Data Conversion Methods," for detailed discussion of various conversion approaches.

Developing the Data Conversion Approach

A variety of issues must be addressed during the development of a comprehensive data conversion approach. Figure 2.4 illustrates the major issues involved. The data conversion approach for a typical GIS implementation project is developed by considering the following factors:

1. Database design, system vendor, and data format requirements (all of which should be documented in a detailed data conversion specification).

2. Sources for each land and facility feature and their attributes, quantities of features, and a detailed database design that includes data relationships.

3. Applicability of available conversion techniques.

4. Project budget compared with expected conversion costs.

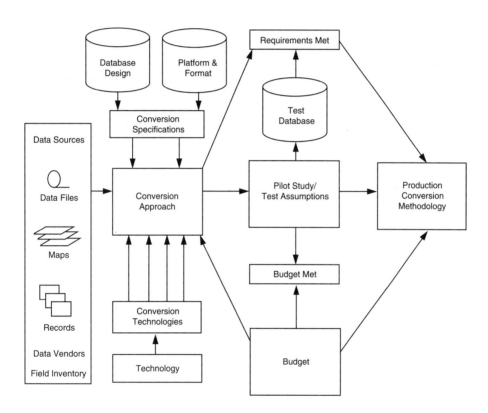

Fig. 2.4. Developing a data conversion approach.

By formulating a plan that addresses each of these factors, an appropriate data conversion approach can be developed. To ensure that the data conversion approach is within the cost estimate, it is common to perform a pilot project (discussed in chapter 12, "Developing a GIS Data Conversion Plan") to test the assumptions associated with the initial conversion approach. The conversion approach must meet data and budgetary requirements.

Developing a data conversion approach may take weeks or months. Time is required to finalize a suitable approach. Due to the complexity and size of GIS projects, conversion approaches can rarely be developed "on the fly." There are many data conversion techniques that may be appropriate; however, the most common characteristic of GIS data conversion projects is that every project is different.

2.3 GIS Data Requirements of the System

The data requirements of a GIS and its attendant applications are the basic set of specifications that will dictate the functionality of the GIS. The compilation of data requirements for a particular GIS is the result of an analysis of user needs. These user requirements are translated into a database design for graphic features, tabular attribute data, and data relationships that support the system functionality desired (see chapter 8, "GIS Database Design"). Figure 2.5 portrays a *topdown* view representing the importance of fully understanding the anticipated GIS applications before embarking upon data conversion.

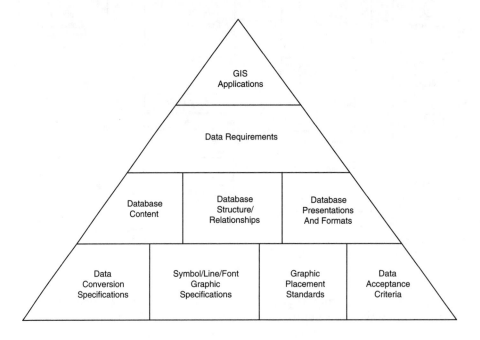

Fig. 2.5. Topdown database design.

Data Relationships

Typically, graphic features and tabular attribute data are arranged according to a series of predefined data relationships, such as an owner name attached to a parcel and/or a parcel attached to a neighborhood area. These data relationships vary in complexity for each GIS project and between GIS products. Some GIS are designed with very simple data relationships, while others are very complex. For example, street names that are related (linked) to streets can provide a basis for locating specific street intersections.

Utility companies may design sophisticated data relationships that support more complex engineering and network design applications. A sophisticated electrical current tracing model will require a comprehensive data relationship that includes the interconnected cables, circuit phases, transformers, fuses, and switches of an electrical distribution network. These types of models are used for analyzing client load on outside plant facilities, for tracking work orders, or for other facilities management applications.

Conversion of graphic features, associated tabular attributes, and data relationships within a spatial context make GIS data conversion a unique challenge.

GIS Vendors and GIS Data Formats

The database feature, attribute, and relationship requirements (database design) must take into account the fact that the GIS implementation must reside and operate on a vendor-specific GIS. In the past, vendor-specific hardware and software formats have deterred the development of universally acceptable data conversion approaches. Recent developments in raster imaging have introduced both new opportunities and new issues in managing data formats. However, the converted data must be loaded to the vendor-specific hardware platform and the selected GIS. The conversion contractor may develop the database using a different GIS than the client will use. In this case, the contractor must reformat or translate the GIS database for delivery to the client's vendor-specific system.

2.4 GIS Source Data

Chapter 4, "GIS Data Sources," covers sources utilized in the GIS data conversion process. The following paragraphs present an overview of GIS source data in the broader context of data conversion.

Generally, GIS are designed and implemented with the intention of automating manual operations. The data used in these manual operations may reside in a centralized location and consist of maps and/or records that are comprehensive, up-to-date, and accurate. However, in most organizations, the required data is fragmented and scattered across a number of documents and records that must be brought together, correlated, corrected, and compiled with respect to a land base. Typically, separate records are manually maintained in different formats and in different departments located in different areas, different buildings, or even different cities. Figure 2.6 shows that, to ensure a resulting GIS database contains the appropriate data, all of the organization's records have to be compiled, reconciled, and interpreted to gain sufficient information to create a meaningful GIS database.

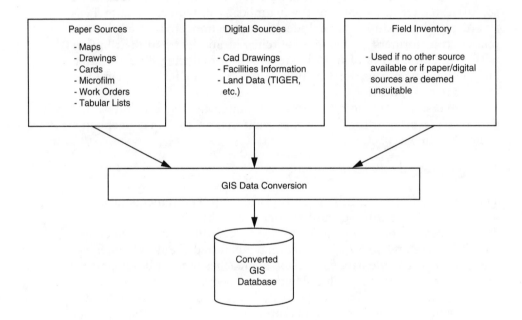

Fig. 2.6. Converting different data sources.

Records reconciliation often requires a substantial amount of time for a large GIS data conversion project due to the significant amount of manual effort involved. In some GIS projects, the absence of accurate paper or digital records requires the verification or acquisition of required GIS data through field inventories.

Geographic information system data conversion normally begins with the identification of the data sources for the land base. These sources may range from extremely accurate surveyed data measured to one hundredth of a foot, to manually maintained maps containing no ground control reference. Facilities data sources must be identified by carefully analyzing a variety of documents, such as facility maps and equipment cards. Source data may also be existing computer files that can be integrated into the GIS database. Geographic information system source data may originate from an organization's own files, from other organizations, or from data vendors.

A prerequisite for GIS data conversion projects is the construction of a source data matrix (see fig. 2.7) that identifies the preferred data source(s) for each feature and attribute found in the GIS database design. If a data source does not exist, the source column for that feature and or attribute(s) should be marked "future" so that the data conversion contractor knows that these features/attributes are not within the scope of the contract. Note that the sample in figure 2.7 is one page from a 22-page source data matrix from an electric utility project.

Document preparation of all the source data is a significant task to be completed before the conversion contractor can begin work. All source data have to be retrieved from storage, verified and updated, cross-referenced with indices, copied, and packaged for delivery to the conversion contractor. Collecting, verifying/updating, organizing, and copying information is a time-consuming task best accomplished by the client organization. It is subject to the prioritization of project areas and therefore can be performed in phases as the data conversion project progresses.

Data preparation also includes improving the clarity of data for people outside the organization who are unfamiliar with internal practices. Many maps and records may need extensive work to make them legible and clear for the conversion contractor's staff. This process is sometimes referred to as *scrubbing* and is often done by the client prior to shipping source data to the conversion contractor. A source data scrub can be time-consuming; ample time should be allowed for this task, as well as for a general conversion plan that prioritizes project areas and data types.

2.5 Techniques Used to Populate a GIS Database

The techniques used to populate a GIS database can be directly related to the complexity and size of the data conversion project. For some small projects, utilization of the GIS data maintenance functionality may be sufficient to support the conversion of graphic features and attribute data. However, in larger GIS data conversion projects, the complexity and size of the data conversion effort requires efficient tools and techniques specific to the data-conversion process.

Entity	Attribute	Street Light Map	OH Inventory Map	UG Inventory Map	Circuit Map	UG Work Orders	Field Work	Field/Existing Records	Aerial Photos
Pole (nonwood joint)			2						1
	Pole Number		1						
	Height		1						
	Year Installed		1						
OH Primary Conductor Segment		1	2						
	Pole #1, #2	1	2						
	Phase A Size	1	2						
	Phase A Material	1	2						
	Phase B Size	1	2						
	Phase B Material	1	2						
	Phase C Size	1	2						
	Phase C Material	1	2						
	Neutral Size	1	2						
	Neutral Material	1	2						
	Number of Phases	1	2						
	Type	1	2						
	Operating Circuit Voltage	1	2						
	Circuit Number	1	2						
	Insulation Type	1	2						
UG Primary Conductor Segment		1		2					
	Size	1		2					
	Material	1		2					
	Number of Phases	1		2					
	Type	1		2					

*Sources are shown either as first (1) or second (2) choices

Fig. 2.7. Source data matrix.

Instead of developing in-house expertise in data conversion, most organizations utilize GIS conversion contractors. Such contractors employ a variety of highly productive, conversion-specific tools and techniques to provide a total conversion service to organizations implementing a GIS (see fig. 2.8). Source data are provided to these contractors, who then perform the appropriate conversion activities and deliver the GIS database in the format required by the GIS data conversion specifications and GIS database design.

Where adequate resources and economies of scale exist, an organization may implement conversion tools and techniques with its own

Fig. 2.8. Workstations at a conversion contractor's facility.

internal staff. With the assistance of GIS data conversion consultants, an organization may make effective use of available staff and GIS hardware resources to accomplish the conversion objectives. An internal data conversion effort usually benefits from the organization's existing knowledge and previous use of the records in day-to-day operations.

The principal techniques or conversion methods are applied according to the type of source data that have to be converted or according to the presence or absence of source maps or records. For instance, maps can be digitized, records key-entered, and parcels precisely calculated, or if new land base data are needed, they can be provided by aerial mapping. Chapter 6, "Data Conversion Methods," examines the techniques of populating a GIS database.

2.6 Making It All Fit Together

Data conversion is a significant process, often complex and time-consuming, that can temporarily increase stress to an organization seeking to implement a GIS. As in other situations where state-of-the-art technology meets a manual records-based culture, data conversion may be quite disruptive. In some cases, the manual-based systems have existed for more than 100 years. Maps that have not left their storage drawers for years are suddenly handled and moved around, and data that were considered reliable and complete may suddenly be seen in a different light.

Data conversion must be carefully coordinated between the conversion group or contractor and the company, as well as among all departments in the organization to minimize potential disruption. Individuals or departments may have to sacrifice some short-term efficiency in order to accomplish data conversion. Those same individuals or departments must also begin to gain familiarity with GIS technology. In larger organizations, the information services (IS) department or other technology providers must also begin adapting to this new system. A successful GIS project implementation requires successful management of the people participating in the conversion effort. In many cases the conversion process involves a reengineering of data handling, maintenance, and use for the implementing departments.

Conversion Management

As we have seen, data conversion is a significant part of implementing a GIS. Issues such as data integrity, accuracy, and user acceptance are additional factors related to the success of the GIS.

Relative to the size of the implementation, the scope of data conversion on a typical project can be said to define a project within a project. Successful project management practices are especially important in such an undertaking. The initial estimates, scheduling, project tracking, production capacity, quality assurance and quality control, training, and change control are but a few components that merit close monitoring in what is still a manually intensive endeavor. The size of many GIS data conversion projects requires commitment to specific roles and responsibilities from a dedicated staff. Figure 2.9 depicts the typical organization of a data conversion project.

Each data conversion project participant is responsible for specific roles and responsibilities. Figure 2.9 shows that they have their particular areas of responsibilities and expertise, including their interdependency. Each project participant has the following areas of responsibility:

Client Executive Committee

- Makes policy decisions pertaining to project.

- Oversees project funding allocations and authorizations.

- Members represent each department/organization involved with the GIS project.

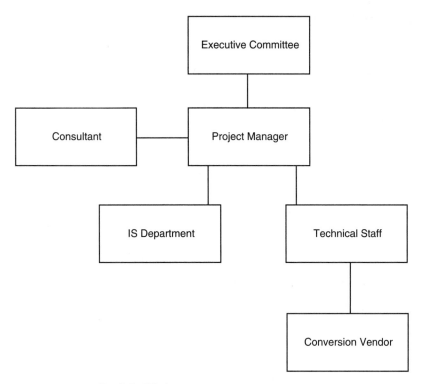

Fig. 2.9. GIS data conversion project organization.

Client Project Manager

- Oversees day-to-day project operations and management.
- Provides technical input and briefings to executive committee.
- Acts as primary point of contact for contractors/vendors.
- Coordinates all activities relating to the conversion project.
- Manages specific budget/schedule responsibility.
- Assumes ultimate responsibility for acceptance of data and distribution of data to end users.

Conversion Contractor

- Develops document preparation (scrub) procedures.
- Scrubs documents (alternate: client/technical staff).
- Conducts maps and records conversion.
- Develops application software.

- Performs field inventories.
- Reports to project manager.
- Develops internal quality control/quality assurance (QC/QA) procedures.

Consultant and/or Conversion Contractor

- Assists project manager/team in developing conversion requirements/specifications.
- Assists project manager/team in evaluating/selecting a qualified conversion contractor.
- Develops project planning strategy.
- Develops QC/QA procedures and acceptance criteria.
- Develops and documents required maintenance procedures.
- Conducts or arranges end user training.

Client Information Systems Department

- Coordinates network communications requirements.
- Performs GIS backup procedures (computer operations).
- Installs GIS hardware, software, and data.
- Develops QC/QA software.
- Develops GIS data maintenance routines.

Client Technical Staff

- Collects source data for data conversion (alternate: conversion contractor).
- Performs document scrub.
- Assists conversion contractor with interpretation of source data.
- Performs QC/QA on conversion deliverables.
- Performs database administration and system management functions.
- Coordinates database maintenance procedures.

Due to the size of some GIS project implementations, conversion prioritization is a necessary part of a conversion approach. Prioritization may result in conversion by specific geographic areas, by feature types, or by other strategies that provide optimum payback scheduling. Incrementally creating a database, or using piecemeal conversion, to satisfy immediate needs has become an attractive alternative when conversion costs prohibit a steady, full-scale data conversion effort. The disadvantage of incremental data conversion is that benefits associated with areawide applications, such as utility network analysis, are delayed. Project management and coordination costs will also be higher.

Other logistical considerations may impact the development of the database. Among these factors are the availability and timing of the use of source data (e.g., the acquisition of aerial photography), managing access to source data being converted, and managing the dynamic nature of the data while conversion is taking place.

A primary factor in determining the rate of data conversion is that the organization will begin using and maintaining the data as soon as conversion is completed for each prioritized area. Given that the just-in-time implementation of converted data may be critical to the success of the GIS project, disruptions to the conversion schedule must be minimized.

Careful thought must go into prioritizing conversion areas.

Risk

The scope of a GIS data conversion project can result in a significant amount of risk to the overall project implementation. A sound approach to data conversion may dramatically reduce the project risk. By carefully evaluating the factors that help ensure a successful conversion effort, by enlisting the assistance of qualified consultants and vendors, and by developing a detailed implementation plan, chances of "making it all fit together" will be greatly increased. The following will ensure a successful GIS implementation:

- Data quality
- Cost containment
- Meeting the schedule
- Communication
- Staff
- Scrub
- Database maintenance upon completion of conversion
- Satisfied users
- Upper management support
- Procedural innovation

Information Services and Data Conversion

A GIS requires a hybrid database to support information management by GIS users requiring spatially oriented data operations. Geographic information systems are relatively new and have evolved dynamically over the past 20 years. In some organizations, the IS department is well suited to play a significant role in the fundamental issues of data conversion and implementation. However, the IS department is often unaware of many of the unique issues related to data conversion for a GIS. It is also rare for an IS department to develop spatially oriented data management expertise on its own.

In most cases, though, it is critical for the IS department to participate on a technical level in system issues and in the use of existing digital data files. As mentioned previously, an organization's internal systems and networks expertise is a decided advantage, especially where an organization has developed a strategic plan for all corporate data, including data maintained via the GIS.

These enterprisewide views most often show the need for IS involvement in GIS implementation and data conversion planning and execution. The true integration of client information, accounting data, stores information, crew dispatching systems, permitting, and many other computerized data systems within an organization will rely heavily upon the expertise of the IS department.

Corporate View of Data

At a strategic level, many organizations have found it desirable to integrate the GIS within a larger, enterprise approach to data management. An organization must carefully design and implement a GIS data conversion approach that not only meets the data requirements of the selected GIS hardware and software but also supports the broader strategic implementation plan. Common keys within a database are developed to provide for data integration, system interfaces, and specific applications that must cross system boundaries. Many common keys are established during data conversion to accommodate these corporate requirements.

Strategic implementation plans minimize confusion.

Chapter 3

Hardware Issues

3.1 Introduction

An important aspect of GIS data conversion is vendor hardware. As software is being developed to satisfy a variety of data conversion needs, one often overlooked aspect is that specialized data conversion software usually depends on hardware components, such as digitizers and scanners, to be of use. As such, hardware and its evolution have influenced GIS data conversion. This chapter discusses the evolution of hardware platforms supporting GIS data conversion.

The GIS data conversion process is evolving as new technologies are tested and implemented. Every year, new solutions are offered, making the GIS data conversion marketplace a rapidly changing arena. All components of GIS are subject to this fast pace, driven by the need to handle geographic-related data in the most cost-effective manner and to convert certain types of data efficiently. As an example, the digitizer tablet was developed to allow the digitization of maps, and the development of the graphic screen was influenced greatly by the need to display high-resolution color images of complex graphic map data.

3.2 Factors Influencing Data Conversion Hardware

Several factors influence data conversion hardware. First, the need to convert certain types of data creates a search for solutions. Second, only proven technology that is available should be used for data conversion. Third, conversion economics dictates what and when hardware can be purchased. And fourth, the data conversion industry experience with specific hardware has to progress beyond the research and development stage so that the solution is used cost-effectively.

The Need to Convert Certain Data

The need to digitize maps initially led to the development of digitizer tablets and then to the development of large document scanners. Also, data conversion can now distribute portions of the process over workstations linked by a network. These solutions are examples of how data handling needs have influenced hardware development and system solutions for GIS data conversion.

Available Technology

Specific types of technology must be available before they can be incorporated into data conversion solutions. For instance, data conversion in the 1960s used card or record-oriented batch-job implementations available then. Conversion projects in the 1970s incorporated the use of CAD systems to produce automated mapping data for projects based on minicomputer-driven terminals. In the 1980s, low-cost and high-performance personal computers (PC) were used to supplement the interactive graphics workstations in the data entry process. Systems in the 1990s are incorporating local area networks and raster image data as part of the conversion process, utilizing sophisticated form-driven user interfaces to capture attribute data efficiently. Raster data from aerial photography and satellite imagery is now a key component of many GIS projects, primarily due to the availability of high performance workstations.

Economics Dictates

Economic forces drive most GIS users to find the most cost-effective methods for conversion projects. However, efforts to reduce conversion costs can discourage the development of new conversion technology. If low price is the primary selection factor for a data conversion contractor, most start-up data conversion companies will be forced to get into their first job as a "loss leader." This is a common strategy for data conversion companies that are also trying to get their first few contracts. The strategy forces most companies to operate constantly on a low margin, which frequently leads to a lack of funds for research and development and for investing in new technologies.

Industry Experience

Because GIS software products are basically different from one another and because each was originally developed on a vendor-specific hardware platform under a unique operating system, no GIS product runs on all platforms. Most GIS run better on one specific platform than on others. If a data conversion contractor wishes to offer services for a variety of GIS, that contractor will need to know how to handle a variety of operating systems. Experience with a vendor-specific GIS and operating system is usually an important conversion contractor selection criterion.

Additionally, a contractor's experience in translating data from one GIS to another is evaluated during the selection process. Often a contractor will convert (i.e., digitize) the data on a system other than that which the client has selected.

3.3 Data Conversion Hardware Evolution

The evolution of hardware used for data conversion has been driven by the need to improve data volume, display, processing speed, conversion, input, output, and editing. For example, large digitizers and plotters were developed to input and output large maps, illustrating how the need to handle certain data has had an influence on hardware development. Considerable increases in available computer memory were needed to accommodate the handling of large data volumes, and this, in part, has helped pave the way for the scanning and display of raster images.

Prior to 1980

The hardware platform for conversion in the 1960s was batch-record oriented, incorporating offline digitizing (no interactive display) to record coordinates, card forms for data entry, and basic batch graphic plotting capabilities. Small computers (such as IBM 1130) were used to create data for specific mainframe applications. In the 1970s, minicomputers appeared, incorporating storage tube graphic displays and digitizing tables to provide interactive graphics systems initially developed for CAD applications. These systems were then used to create data for similar drawing-based AM systems. Storage tube graphic displays were also added to mainframes for FM systems emphasizing database applications with graphics as a secondary output form. However, purchase and operating costs prohibited using the mainframes for data conversion.

The 1980s

During the 1980s, data conversion processes began using high-speed interactive graphics workstations with raster displays first introduced by CAD vendors. These workstations were used in conjunction with large minicomputers. Organizations recognized the importance of tabular database applications in supporting business-critical applications, the need for topological data structures for spatial analysis applications, and the need for high resolution graphic display screens to inspect the results. Geographic information system data conversion techniques had to keep pace with these recognitions. Vendors of CAD started adding attribute data to the graphic feature capabilities of their systems. This added capability was generally accomplished by using proprietary technology since few database management systems were available for the CAD operating systems and platforms. At the same time, numerous stand-alone PC-based topological systems with spatial analysis capabilities began to be used successfully in small organizations. The PC systems also had to add increasingly sophisticated graphic and database functionality.

Meanwhile, mainframe vendors, with their knowledge of corporate database environments, added many graphic functions to mainframe capabilities, thereby integrating attributes and graphics functionality within the same database management system. However, it was not economically feasible for mainframes to be used for small data conversion projects. Most GIS data conversion projects utilized CAD-based conversion approaches because of their lower initial cost and higher performance on smaller jobs. But data conversion companies found that poor conversion performance outweighed the initial advantage of inexpensive PC GIS hardware. Therefore, some of the larger GIS projects used a mainframe system for data conversion as well as for data maintenance.

During the 1980s, organizations purchased GIS products on the basis of available budget monies, functionality, and/or database type. These organizations then approached GIS data conversion contractors to obtain conversion services for their vendor-specific GIS. Because all GIS products were proprietary in nature and ran only on one specific hardware platform, conversion contractors faced the prospect of having to purchase a variety of GIS hardware platforms in order to satisfy a variety of clients. This, of course, was hard to justify, given the low profit margins of the data conversion business. A generic conversion platform was very difficult to implement due to the inherent incompatibilities among the vendor-specific GIS products. This is one force behind the adoption of "open systems architectures" now being pursued by GIS vendors and hardware developers.

Ongoing hardware evolution provided additional support for the GIS data conversion process. High-speed electrostatic plotters opened a conversion bottleneck associated with the creation of the many "check plots" required as quality checks for delivered GIS databases. Workstations became truly stand-alone, providing large local central processing unit (CPU) memory and disk storage sizes. Conversion contractors used more powerful PCs for attribute data entry and plot generation. Color technology was used to assist the input and editing of complex GIS graphic data.

Scanners and raster technology were introduced as part of the conversion process.

While the PC had been available early in the decade, only in the late 1980s did it play a significant role in GIS data conversion processes. Not until the general availability of the Intel 80286 and 80386 based systems was there adequate processing power and network functionality to support data conversion activities. The PC became an effective alternative to mainframe and minicomputer connected terminals.

Reduced Instruction Set Computing (RISC) workstation technology also became available near the end of the decade. This technology offered tremendous price/performance advantages over earlier GIS hardware platforms.

The concept of *client-server* computing evolved in conjunction with open systems. Client-server computing divides an application into two or more modules, each residing on a separate computer processor. Each module has streamlined access to the resources best suited to fulfill its function. For example, database modules run on file servers that manage disk drives, while the client process modules run on a workstation with a graphic display device to manage the user interface of the application.

Workstation operating systems for GIS became primarily UNIX based. Several different relational database management systems emerged that were based on Standard Query Language (SQL, sometimes pronounced *sequel*), a widely accepted query language. Other standards such as the X Window System and the TCP/IP network protocols also emerged (see fig. 3.1).

The 1990s

The introduction of image-based technology, scanners, and low-cost optical disk storage helps reduce the GIS data conversion cost barrier. All major GIS support the use of image data. *Incremental conversion* allows users to convert maps to raster images through scanning, and to store the images on a raster image GIS. The conversion of raster format data to vector format data is done only in geographic areas or GIS data layers that are subject to certain kinds of analysis. Rapid, low-cost map scanning, together with operator assisted vectorizing of a displayed raster map image, is called *heads-up digitizing*. This process allows GIS users to support internal GIS data conversion environments that can reduce or delay up-front data conversion costs.

Reduced Instruction Set Computing (RISC) technology continues to offer improved price to performance ratios. Computing capabilities previously available only on mainframe computers are becoming available on RISC platforms, making it possible to do mainframe GIS data

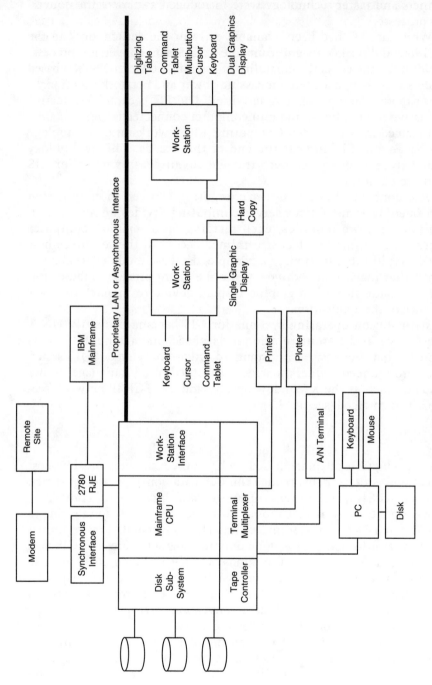

Fig. 3.1. Typical hardware configurations in the 1970s to early 1980s.

conversion on workstations. Integration of GIS database management with other enterprise database management systems provides organizations with access to corporate geographic-related data throughout the enterprise. Switched, high-speed digital communication networks are providing low-cost, on-demand access to volumes of GIS and raster image data from widely dispersed locations.

X terminal technology is being used in organizations for low-cost queries and occasional use applications, particularly in GIS configurations with a large number of users. Data conversion contractors may use X terminals for GIS attribute data input and/or for data validation/checking processes. The use of X terminals for data conversion, however, depends on the X terminal configuration, providing improved price and performance over a PC (see fig 3.2).

Artificial intelligence expert systems technology is being used in an incremental attempt to provide automatic GIS data conversion of raster image data to intelligent vector databases. Although there are a few successful GIS data conversion expert systems available, this approach will not be effective for general GIS data conversion until a new generation of computers becomes available. These new computers will better support expert system processing through special functions in their hardware and operating systems.

3.4 In-House (Internal) vs. External Conversion Hardware Issues

Conversion contractors can offer organizations rapid start-up of GIS data conversion due to their experience with similar GIS projects and a standing investment in conversion technology and tools. First-time in-house GIS data conversion projects must face the challenges of acquiring staff and training, developing conversion procedures and techniques, and installing sufficient GIS hardware (which may become excess after completing the data conversion process). Data conversion done in-house may also be subject to internal influences such as budget adjustments and changes in priorities. Also, in-house data conversion projects often underestimate the total investment and hidden costs of this part of GIS implementation and may be more lax in quality control than an external conversion contractor. In-house conversion processes require the company to assume the overhead of systems administration, software support, hardware maintenance, and computer operations, whereas the conversion contractor performs these processes as part of the contract price and the client is never directly exposed to these inherent costs.

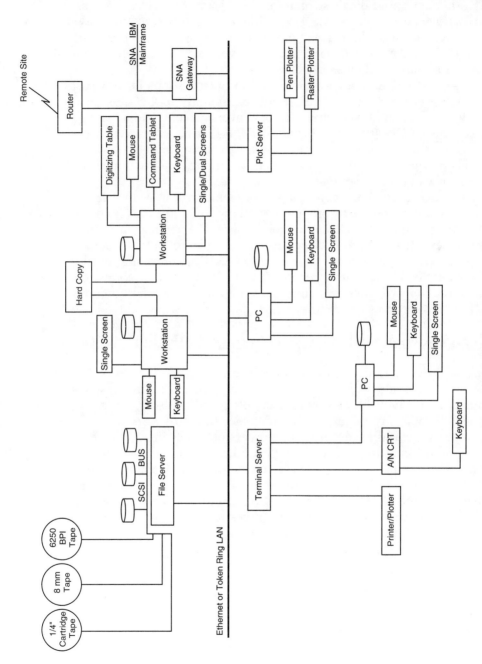

Fig. 3.2. Typical hardware configuration in the late 1980s to 1990s.

3.5 Input Hardware

Input hardware are the devices used to enter data into a GIS database. Other than for scanning, input hardware is independent of the data source and usually involves operator activity. For example, a conversion contractor operator may be reading a written or printed record while entering data through a keyboard. Apart from information that is read or interpreted by an operator, graphic feature position or shape information is entered with the help of a digitizer or a scanner.

Keyboard

The traditional keyboard is still an indispensable input device for all types of GIS data conversion, separate from or in conjunction with digitizing activities. Desirable keyboard features include standard layouts for numeric keypads, cursor control, and user programmable function keys. These features allow keyboard-trained clerical personnel to efficiently enter large amounts of tabular attribute data into a database. A keyboard is now supplemented with a mouse or cursor to select from a list of standard values.

For field inventory operations, the trend is toward using hand-held data collectors (a type of keyboard) to directly record the information collected in digital form.

Command Tablet

A command tablet is a small digitizing tablet normally overlaid with a menu of commands used for rapid command selection. The tablet can also be used for tracing sketches at a lower accuracy level than with a digitizing tablet. Command tablets are beginning to be replaced by on-screen menus due to X Window System technology that allows simultaneous displays of graphics, images, and text (tabular) data.

Graphic Input Device

Virtually all GIS have an input device that controls a cursor on the graphic display screen. This is usually a low-cost, low-accuracy pointing device such as a pen, mouse, track ball, or joy stick. Touch sensitive screens are available but are not commonly used in GIS applications, with the exception of public access terminals. Graphic input devices are used to select command options from menus and forms presented by a graphic user interface (GUI), to select graphic features (entities) displayed on a screen, and to perform some graphic editing functions on the screen.

Digitizing Tablets

Although their usage is being diminished by scanning and heads-up digitizing, digitizing tablets are still used to convert existing maps and drawings into GIS format. The operator places a graphic input device on a point on the map, and the digitizer sends the tablet position of the point to the computer, which in turn calculates the x,y ground coordinate position for the selected point. Digitizing tablets typically support E-size and larger drawings, with 0.001-foot resolution, 0.001-foot repeatability, and 0.005-foot accuracy.

Scanners

Scanners convert existing maps and drawings into a digital raster format. When scanning a document, rows of pixels or dots record a value representing the gray scale intensity or color of each of these dots. Scanners with varying resolutions of 200 to 1,000 dots per inch (dpi) are typically used for scan conversion of documents and maps. The scanning process can be done in one of three modes: black and white, gray scale, or color. The type and condition of source documents determine the resolution and mode selected.

Pen-Based Laptop/Hand-Held PCs

Pen-based laptop or hand-held computers are beginning to be used by GIS data conversion contractors as a data collection tool for field survey and verification work.

3.6 Output Hardware

Output hardware is an important aspect of the GIS data conversion process because some parts of the process do not require direct access to a workstation. For example, a major quality control task in the data conversion process is to plot a hard copy of the GIS data and to perform a visual edit to verify that all of the geographic data found on the source documents are in the GIS. This process entails a one-to-one check of all conversion source data to the corresponding GIS symbols/features shown on the plot.

Pen Plotters

Pen plotters are low-cost, low-speed plotters, capable of drawing on paper, mylar, and vellum using liquid ink, ball point, felt tip, and other types of pens. Pen plotters generally support multiple pen capabilities for colors and line weights. They are capable of producing maps/drawings up to E-size, and have maximum drawing speeds of about 30 inches per second. Pen acceleration rates vary, with slower accelerations tending to be more accurate. Pen plotters also vary greatly in accuracy, performance, and cost. Accuracy requirements are typically 0.001-foot accuracy and 0.001-foot repeatability. Pen plotters are often preferred for the production of urban maps of dense utility infrastructure information. They are also used to produce presentation quality plots, except for area shading/choropleth maps, which are better produced on raster plotters.

Raster Plotters

Raster plotters use either electrostatic or laser xerographic processes to create plots at resolutions between 200 to 800 dpi in either black and white or color. Color raster plotting is substantially more expensive than monochromatic plotting. Desirable raster plotter features include hardware vector to raster conversion, raster image plotting, and raster with vector overlay plotting.

Typical electrostatic plotters produce output up to E-size at a resolution of 400 dpi, but with relatively low accuracy (1 percent of paper length) due to paper stretch and roller feed. Typical laser xerographic plotters are smaller format (B-size, or 11 inches x 17 inches) with 300 dpi resolution. Laser plotters do offer various operational advantages over electrostatic plotters, such as plain paper output and fewer chemical problems.

Screen Copy Devices

Screen copy devices are used to obtain a paper copy of a workstation screen display without going through the plot creation process. The time needed to produce a screen copy is usually on the order of a minute or so. Screen copies are created when the GIS workstation sends black-and-white or color output to a small-format raster plotter (electrostatic, laser, or thermal wax transfer). Screen copy devices may be connected directly to a GIS workstation, multiplexed to several GIS workstations (usually up to four), or connected to a network of many workstations. Screen copy devices are generally used to obtain workprints or checkprints of specific areas of a GIS database. They greatly aid communications during the conversion process, both within the data conversion team and between the conversion contractor and the client.

Computer FAX

Facsimile (FAX) has long been a tool to support communications between different physical locations. Conversion contractors and their clients rely on FAX to communicate clearly and quickly and also to provide a record of the communications. Computer FAX can be used for sending and receiving small plots and memos to and from standard FAX machines. This process is quicker and provides higher resolution quality than first plotting to paper and then sending manually by facsimile. Similarly, FAX transmissions can be received and stored as raster images to be viewed or plotted later. Computer FAX are available as single-user or network-server models. Also, large format FAX machines are commercially available.

Printers

Traditional printers are required during data conversion to output tabular GIS database reports. These reports are used to audit/verify the accuracy and integrity of the GIS attribute data. The printers can be line printers, dot matrix, laser, or electrostatic. Dual function printers/plotters often serve as screen copy devices as well.

3.7 GIS Data Conversion Workstations

A typical GIS data conversion workstation consists of a base system unit containing a microprocessor, random access memory (RAM), hard disk, and video display controller.

Performance levels of workstations used for GIS data conversion purposes are, in most cases, at the lower end of the range available. Workstation options to support three-dimensional operations and area shading are generally not desirable from an economic standpoint; however, image processors to enhance raster operations (such as pixel block moves for menu and window operations) are desirable due to the increasing use of scanning and vector/raster overlay in data conversion.

Mid-range processor speeds (excluding image processors) generally suffice for data conversion GIS workstations since sophisticated analysis and display operations are usually performed by end users of GIS data rather than by the creators of such data. Memory requirements are similarly modest, except when raster image processing/manipulation is involved.

Color raster displays are the norm for today's GIS workstations. Workstation screen displays need to have high resolutions (near 1,000 x 1,000 pixels) on large monitors (19-inch diagonal and larger). The number of simultaneously displayed colors should be 32 or more for data conversion purposes. Dual graphic screen display GIS workstations are commonly used by data conversion contractors for processes that require intensive graphic display functions.

Workstation ergonomics are very important in data conversion since the operators of these workstations operate them for extended periods of time. Desired ergonomic features include height and rotation adjustments, right-handed or left-handed operation, low-reflectance screens, and movable keyboards and command tablets. However, conversion contractors often emphasize price-performance ratios over ergonomics.

The diskless GIS workstation, usually called a *terminal*, is an atypical configuration for data conversion since it completely depends on a remote computer and data storage for all its operations. A failure or problem with the remote computer and/or data storage significantly impacts data conversion production due to the multiple terminals affected.

Geographic information system workstations are usually categorized into four types.

Digitizing Workstations

This is a high-performance GIS workstation with a large precision digitizing table and a high-resolution display with single, dual, or virtual-dual graphics screens. This type of GIS workstation typically has higher performance microprocessors and image processors. Some types of GIS data conversion require the use of these workstations for high volume graphic data input and/or raster image processing.

Review/Edit Workstations

A medium- to high-performance workstation with a single, dual, or virtual-dual graphics screen and a small digitizer tablet typifies this group. These workstations can perform all the functions of the higher-performance workstations except for the digitizing of large maps and drawings (on a single, large, digitizing surface). A review/edit workstation is typically used for reviewing previously digitized data and for query, analysis, and design functions. These workstations are also often used to support the QC/QA functions of GIS data conversion.

Review/Tabular Attribute Data Input Workstations

These workstations offer moderate performance with a single graphic display screen used to browse, display, and review graphic features and tabular attribute data. This type of workstation is often heavily used in the data conversion process for data entry.

X Terminals

X terminals are workstation-like devices that provide graphic display and input capabilities in a multiwindow environment using the X Window System communications protocol. An X terminal requires a host computer (either a workstation's or client's) to actually run the processes that interact with each window on the X terminal. Functionally, X terminals are equivalent to the lower-performance workstations discussed above. As mentioned earlier, X terminals are not preferred for data conversion due to the potential impact that results when a problem occurs with the host computer or disk storage.

3.8 Storage Devices

Storage devices are used for three principal GIS functions: holding data that are being converted, transporting data during delivery, and storing data at the client's site. The capacity of storage devices used by conversion contractors is usually large, allowing several projects to be actively worked on. Data is transported on storage media that use widely accepted formats, such as nine-track tapes and cartridge tapes that are readable on devices based on the Small Computer Systems Interface bus (SCSI). Data conversion contractors and their clients utilize a variety of storage devices, ranging from small diskettes to large storage units such as removable disk drives and *jukeboxes* for optical disks.

Magnetic Disk

Magnetic disk storage is the main storage medium for GIS data during the data conversion process. It provides storage for the conversion contractor's proprietary data conversion software and GIS software as well as for graphic and attribute data during the input, edit, correction, and verification phases of conversion.

Disk storage is provided within the workstation on a network-based file server or a minicomputer (diskless operation) or both. The capacity of individual workstation disk storage units is generally large in order to support the local handling of data. Large memory capabilities allow most workstations to operate as stand-alone units or as data-sharing workstations while minimizing geographic data exchange over a network.

Performance of the disk subsystems is a major factor in the overall performance of the workstation. Network-based disk storage presents a potential bottleneck for operations involving large volumes of data, especially raster image manipulation.

Interfaces to peripherals such as magnetic disk, optical disk, magnetic tape, and scanners are becoming standardized. The most prevalent and therefore convenient interface used is the SCSI.

Optical Disk

Optical disk storage provides an inexpensive alternative to on-line disk storage under certain circumstances. If the limitations of the optical disk technology, such as access speed, do not severely impact the performance of a particular conversion process, optical disk provides a very convenient means to store large quantities of data. For example, the storage of a large quantity of rasterized maps, in which one scanned E-size map can become a file of 10 to 20 megabytes, can be handled through document management technologies.

Some limitations of optical disk storage include slow data access times and slow data transfer rates. Optical disks usually are read only or Write Once, Read Many (WORM) times. Not all applications can operate with media of these characteristics. Erasable optical disks are available, but again, input/output (I/O) performance is limited. The most common application of optical disk storage to data conversion is to archive converted data for future reference and, in some rare cases, for system backup purposes. The following describes each optical medium in greater detail.

1. *Compact-disk, read-only memory* (CD-ROM). Produced through a separate mastering process and suitable for applications where multiple copies of a single large data set need to be produced and distributed. To date CD-ROM has not been utilized extensively in data conversion.

2. *WORM*. A more recent technology that provides for the direct creation of read-only optical disks. Since the disks are removable, this provides a very good archive for data. The performance of this system in write mode is much slower than in read mode.

3. *Erasable optical disks*. Allow multiple writing of the same physical media. They typically have somewhat lower performance characteristics than the other optical media both in terms of capacity and access speeds.

Optical disks may become an integral part of data conversion hardware when performance drawbacks are eliminated.

Magnetic Tape

Magnetic tape is still the most prevalent medium for both data backup purposes and data delivery between conversion systems and target systems. There are three principal magnetic tape formats.

1. Eight-millimeter magnetic tape systems use video cassettes as the storage medium and are capable of storing 5 gigabytes or more of data on a single cartridge. This provides an ideal mechanism for data backup and for delivery of very large data sets. This is becoming a standard medium for delivery of converted GIS databases.

2. One-fourth-inch cartridge tape units are used on a majority of graphic workstations and by equipment vendors for software distribution purposes. These devices offer medium capacity (in the range of 160 megabytes per cartridge) and are suitable for data transfer of smaller data sets.

3. One-half-inch reel to reel tapes at 1,600 bits per inch and 6,250 bits per inch are not used as commonly as they have been in the past. Large physical size and medium performance levels have resulted in reduced use of this storage medium. However, they are still used when one of the systems involved in a data transfer is a mainframe or minicomputer.

3.9 Processors

Processor technology has been advancing at a tremendous pace especially in the area of RISC processors. Raw processing speeds for the microprocessors in workstations and servers easily exceed those for minicomputers and many mainframe class machines. Random access memory capacities have similarly increased (16 to 32 megabytes of RAM on a single board is not uncommon) to allow stand-alone and networked workstations to hold the operating system, the GIS application software, and a large portion of a geographic database in memory. Internal system bus speeds have also become substantially faster to match the higher performances of internal and peripheral components. This increased performance allows faster display of large quantities of graphic data, such as the display of a map. In addition, the increased performance allows scanned raster data to be used as an integral data conversion process, including the now widely used heads-up digitizing approach used by many conversion contractors. Virtually all processors now have 32-bit-wide address and data paths, integral double precision floating point processors, and virtual memory management.

Workstations

A workstation is the primary mechanism by which a user in a data conversion environment interacts with the GIS database. Normally the workstation processor is dedicated to supporting a single user and is involved with displaying data, menus, and messages. While some background processing may occur, this is still relatively infrequent. Certain data conversion activities may migrate in the future to X terminals that are supported by a server class workstation (*client* in X terminology).

Servers

In general computing environments there are many types of servers, but in the data conversion environment a server normally refers to a file server that manages the data and databases that are part of the data conversion effort. The server typically manages a large number of disk drives through a high-performance intelligent disk interface that provides optimization of access requests. In addition, servers often perform computational or I/O intensive tasks such as database verifications, data backup/recovery, and plot processing. In addition, the file server often supports the server portion of a database management system to which individual workstations function as clients.

However, the terminology in an X environment is reversed. A workstation or host computer storing data and/or programs utilized by another computer (i.e., a PC or another workstation) is referred to as the *client machine* in X terminology. The X terminal itself is referred to as the *server* (X-server).

Minicomputers

Until the late 1980s, minicomputers were the nuclei of data conversion systems, and many installations still use them in this way. These systems host alphanumeric terminals and older workstations while also acting as file servers for new classes of workstations. The majority of these minicomputers significantly lack the processor speed, main memory capacity, and disk system I/O performance that are currently state of the art. In installations that have been doing data conversion for some time, the existing body of conversion software and operating procedures make abandoning these computers difficult even in light of their near obsolescence.

Mainframe Computers

Data conversion contractors generally find it difficult to justify the purchase of a mainframe computer. The need for a mainframe in data conversion is rare. Data conversion may require considerable storage space, but seldom requires a large computational capacity exceeding that of modern workstations. On occasion, the customer's mainframe is used by conversion contractors for on-line data conversion.

3.10 Communications

Workstation communications over a network are usually a necessary part of GIS data conversion because of the large amount of data that has to be converted. In addition, the size of the resulting databases is usually greater than the size of any one of the localized disk drives. Communications hardware in a GIS conversion operation generally consists of three types.

1. Terminal connectivity is required to support alphanumeric terminals, PC, or workstations that communicate through serial ports. Data transmission rates for these types of connections range from 2,400 baud to 19.2 kilobaud.

2. Network connectivity is required to support workstations, file servers, X terminals, etc., that operate over local area networks such as Ethernet and Token Ring. Local area networks operate at data transmission rates of 10 megabits per second on Ethernet or 16 megabits per second on Token Ring.

3. Long-distance communications that form a wide area network may be required when on-line access between sites in geographically separate locations is necessary, such as when a data conversion site is remote from the end-user installation and real-time access is required between the systems. Long-distance data transmission rates vary from 2,400 baud dial up access to 1.544-megabit-per-second dedicated T1 channels.

Where larger enterprises are developing long-range strategic plans for large scale integration of computer systems through standardized open networks, GIS data conversion companies are motivated primarily by short-term benefits. Connectivity decisions are driven by the need to get data from one vendor-specific GIS to another in an efficient manner. In the long run, GIS data conversion systems will benefit as much as other enterprises from the application of standard network protocols allowing the interoperability of dissimilar hardware and software systems.

Conversion contractors stand to gain much as a result of the computer industry's move to open systems, even though they generally are not able to develop long-range plans for migrating to these systems. Their decisions and funding are directly driven by the nature of the GIS data conversion projects they are able to acquire.

3.11 Survey Input

An important component of GIS data conversion hardware is not usually thought of as conventional computer technology. This component consists of a variety of data collection and conversion tools that have recently been computerized. These tools are used for such tasks as aerial or field surveys and can directly create digital files that can be translated into specific GIS formats. Related methods are presented in chapter 6, "Data Conversion Methods."

Photogrammetric Devices

Digital and analog stereoplotters are specialized types of GIS data collection devices. They are used to map the geography of terrain for the creation of a land base. Recently, these devices have been coupled with a variety of CAD workstations and, in some cases, directly with a GIS workstation. Stereoplotters do require a very high level of specialized personnel skill and a very knowledgeable management team.

One of the photogrammetric services performed by aerial mapping companies is the preparation of orthophotos and, more recently, raster or digital orthophoto images. The purpose of the orthophoto process is to correct for the image distortions that are caused by the terrain relief. For instance, a straight section of highway running over a hill will appear bent on aerial photographs, and the orthophoto process straightens out the image of the highway. These processes are computer supported, and the preparation of orthophotos or raster orthophoto images requires highly specialized processor-based equipment. The production of corrected images can be part of the GIS data conversion process, providing either photographs that are digitized or raster images that are directly used as a displayable GIS land base.

Total Stations

Total stations are field survey devices that record measurements, such as distances and angles; or calculated values, such as point coordinates; on portable processors with memory. The operator of this device locates features through an optical telescope and then, through a keypad, enters a feature code. This tool can be used to collect the positions of objects that are hard to see on aerial photographs or that are not represented on any conversion source. Examples of such objects include poles, valves, manholes, signs, and natural feature points.

GPS

Aerial mapping companies are beginning to use Global Positioning System (GPS) devices to provide a direct means of determining latitude, longitude, and elevation coordinate values from the measurement of satellite positions. These devices receive signals from several NAVSTAR satellites and compute coordinate positional information based on the signals, with the help of portable or in-office processors and software. Some of the devices allow storage of coordinates of measured points, which can be loaded directly into a GIS. The potential of a GPS to lower the cost of building a GIS land base makes it attractive to data conversion contractors. The capability of determining and storing geographic location information can be particularly useful for inputting facility locations during a field inventory process.

3.12 Field Data Entry Stations

GIS data conversion projects often require that some field verification of data be entered into a GIS, or a field inventory may be an integral part of a particular data conversion activity. Portable and laptop computers provide the capability to do data entry away from the normal office environment where data conversion normally takes place. The normal attribute data entry functions performed on an alphanumeric terminal workstation can easily be performed with a portable computer in field locations. Current resolution capabilities of portable computers provide the functionality for displaying graphic data and even offer some amount of graphic editing in the field. But the principal use of these tools is the creation or review of tabular database attributes in the field, such as transformer types and identification numbers, pole numbers, etc. These new pen-based systems are currently being implemented as data collection devices for several data conversion projects.

Chapter 4

GIS Data Sources

4.1 Introduction

This chapter focuses on source materials, such as maps and records, that provide the basis for the data in the GIS database. The chapter also relates different sources with the types of organizations that typically possess or utilize them.

The term source is very significant throughout this discussion. Geographic information systems are usually not delivered with data, so there must be a source for each and every byte of information that is ultimately stored on the system.

Webster defines *source* as follows: "A point of origin. A record, as a book or document, supplying primary or firsthand information." This applies directly to GIS data conversion where there are multiple, varied, and sometimes duplicate data sources. The sources for the conversion process serve as the primary information used to initially populate the database. Therefore, a prerequisite to commencing the data conversion process is selecting an appropriate source for each item of information to be stored in the database.

4.2 Key Issues Involving Data Sources

One of the biggest factors influencing the degree of complexity of a GIS data conversion project is the quality and coverage of the source data. An analysis of the quality and coverage of possible sources is required because a particular GIS feature is often shown on multiple sources of varying quality.

The quality of source materials must be equal to or greater than the desired quality of the converted database; otherwise the source should be rejected and replaced. For example, if maps do not show the required information with appropriate accuracy, precise calculations or aerial mapping may have to be used to provide the data. This underlines the importance of identifying the quality of sources as part of database definition, before the cost of conversion is estimated and quoted, and data conversion begun.

The parameters that follow describe the quality and coverage of source material; they are similar to the ones used to assess the quality and coverage of the database itself.

- Accuracy
- Coverage
- Completeness
- Timeliness
- Correctness
- Credibility
- Validity
- Reliability
- Convenience
- Condition
- Readability
- Precedence
- Maintainability

Accuracy

The positional accuracy of source maps or drawings must be determined or verified. The results of this analysis will show whether or not the map or drawing is of sufficient positional accuracy to support the envisioned GIS applications. In general, considering that data conversion costs increase proportionally with required accuracy, data needs of specific applications should drive the accuracy needs of the database. If the sources are not of the quality necessary to support certain applications, it may be necessary either to change or cancel the applications or to find better sources.

An analysis of the accuracy of a data source must also consider additional error introduced during the data conversion process itself. For example, a digitizer operator will usually miss the precise mathematical center of a line during the digitizing process, and this error will be reflected in the database. Therefore, the combined effect of source error and data conversion error must be considered during the determination of source accuracy levels and the establishment of quality control error thresholds.

Coverage

The geographic coverage of the source document should be inventoried and compared with the required mapping extent of the GIS project. Often the geographic coverage of sources does not cover the entire GIS project, or there may be holes in the coverage. Typically, other data sources, including field inventory, must be added to complete the intended coverage.

When multiple sources with different accuracies are used to complete a database's coverage during data conversion, the user must be alerted to what accuracies can be reasonably expected within various portions of the database. For example, a multiparticipant GIS project may include the city and county. The city may have current large scale aerial photography to use as a data source for roads, while the county may only have small scale USGS quad sheets that were enlarged for their mapping needs. The accuracy of the data feature roads will be very different inside and outside the city limits.

Completeness

Sources should show most, if not all, specific features. For example, roads should not be missing, and all roads shown should have a name attached. If the source is not complete, then it must be completed in the scrubbing process. On one source, "West Ninth Street" may be spelled out; on another, "9th" may be all that appears. The conversion specifications in such a case must clearly indicate the preferred source from which street names are to be taken. Source precedence must be determined for each and every item to be converted into the GIS database. Secondary and tertiary sources should be provided to cover instances where a data item is missing from, or illegible on, the preferred (primary) source. The secondary or tertiary source may be field inventory when other acceptable data sources do not exist.

However, the inclusion of secondary and other sources in the design of a conversion project often increases cost due to the need for additional document handling, and the number of questions raised because of conflicts among sources.

Timeliness

The timeliness of each source document is another factor that is reviewed during the source evaluation process. Timeliness is also referred to as *currentness* or *vintage*, since it represents how current the information contained on the source map or document is. A source type that is one to two months current will usually be preferable to one that is only updated annually.

On many GIS projects, the currentness of the information is vital. For example, should a school district want to optimize bus routes, such routing should utilize the most current maps. When evaluating the timeliness of a data source, the user must note that all items in the source may not be updated at the same time and there may be multiple update cycles for a particular source. Each feature should be evaluated individually to determine the timeliness of the information. Some utility companies indicate that they update their maps every two weeks with new facilities information, but the property lines that are shown on the map may be updated on a annual basis only, or not at all.

Correctness

The correctness of a data source measures whether it shows information that matches the real world. For example, a highway is shown as such, and not as a river, and vice versa.

Correctness is a crucial factor for GIS data conversion. At times it may be necessary to use field inventory methods to make sure that the data sources are correct. For example, design personnel of an electric distribution company have to know that a certain business is connected to a specific type of transformer as shown on a design detail and that the crew did not install another more conveniently available transformer instead.

Credibility

Source type credibility is also considered as a data quality factor. Often, credibility is established through *synchronization* among multiple source types. If multiple sources agree on the value of a certain attribute, credibility in that value increases. In cases of conflict, recently field-verified sources will usually prevail over outside plant accounting records. In general, personnel who have to use a certain source will be able to judge its level of credibility.

Validity

Validity is another factor to be assessed. The question asked for each data item on each source type is: "Does this source contain only valid values?" For example, if a utility purchases and installs poles that are available only in heights of 5-foot increments, a pole shown as 37 feet is not valid. This incorrect value could reduce the overall validity of the source.

Correctness differs from validity in that it indicates how well the information depicted upon the source actually reflects the real-world (as-built) conditions. Using the example above, if a source depicts a 35-foot wooden pole on the northwest corner of Sixth Avenue and Sheridan Boulevard, it is considered valid since 35 is divisible by 5. However, if there is really a 40-foot wooden pole physically residing at that location, the source information is deemed incorrect.

Reliability

Over time an organization tends to develop an informal opinion as to a particular source document's reliability. If field crews and engineers continually encounter field conditions that are dramatically different from what is shown on their operating maps, those operating maps will be considered an unreliable source of information. The cause of the unreliability may be a variety of reasons, but regardless of this, the map users have simply confirmed that the information contained on such maps cannot be relied upon.

Convenience

Convenience is often an important factor in assessing data source suitability for conversion purposes. This factor merely deals with how easy it is to locate and use the source for data conversion. However, logistical complications often arise. Usually the more convenient a source is, the more it will be required for continued use by the organization's personnel during the period of data conversion. Therefore, reproduction of all source documents must be evaluated and possibly planned for.

Condition

Many source documents are so old and frail that special handling and reproduction techniques must be employed to minimize further document deterioration. In any case, the information on maps and records is much more important than the paper it is written on, and it can be lifted photographically or electronically (scanning).

In some cases it may be necessary to change the source by transferring it to another medium. This can be in the form of laminating the old map or record, copying the map or record onto a stable material, or scanning the material so that a raster image is stored on magnetic media. Generally, this is not a data conversion process per se but sometimes is provided as part of data conversion services. In the case of scanning a source, automated or software-supported data conversion can be used to convert the raster images.

Readability

Legibility or readability of source material is perhaps the most important of the source quality factors because data conversion is usually done by contract personnel who are not familiar with the sources. The items specified as originating on a particular source must be readily and consistently legible on that source type. Consultants often caution clients that if the client's staff cannot read the source, the conversion contractor surely will not be able to read it.

Unfortunately, the reproduction process often reduces the overall legibility of a given source type. A legibility analysis should be conducted on test versions of the sources (copies or blueprints) that will be used in the data conversion process.

Precedence

As a rule, source material will show duplicated detail. For example, utility maps will show land information such as streets, but this information has been copied from other maps. During the copy process, the land information will have certainly suffered somewhat in quality; therefore, it is usually preferable to utilize the *preceding* source. But this is not always the rule because sometimes second or third generation maps have been maintained better than the original source.

Maintainability

In the context of data conversion, the concept of data maintainability takes on a slightly different meaning. It is common for the GIS database, once it is constructed, to replace all or many of the sources used to produce it. Accordingly, the ease of maintenance (or lack thereof) that was historically associated with a particular source document becomes academic. Once the GIS is fully implemented at the client's site(s), the maintainability concern shifts to the GIS. New procedures are generally required to assure that any maintenance difficulties that were present in the pre-GIS era are not perpetuated after the GIS is installed.

4.3 Maps

Most organizations implementing a GIS already use a large number of maps. Virtually every GIS data conversion project entails the use of hard-copy maps as source documents. These maps vary widely in terms of content, media, format, age, scale, complexity, and usage (i.e., the number of departments within the organization that utilize a particular map type). Tables 4.1, 4.2, and 4.3 illustrate the maps various organizations use, in terms of types of maps, general map content, and principal use. (The shaded areas in tables 4.1 through 4.3 denote an organization's map requirements.)

Table 4.1. Map type.

Organization/Department	Map Type							
	Electric Facility	Water	Sewer	Gas	Property	Topographic	Planimetric	Telephone Facility
Electric Utility	■						■	
Gas Company				■				
Water Company		■	■					
City/County - Public Works		■	■			■	■	
City/County - Planning					■		■	
City/County - Tax Assessment					■			
City/County - Police/Fire					■		■	
Telephone Company								■

Table 4.2. Map content.

Organization/Department	Map Content										
	Poles	Substations	Transmission Lines	Water Mains	Sewer Services	Parcels	Subdivisions	Contours	Elevations	Streets	Water
Electric Utility	■	■			■					■	
Gas Company										■	
Water Company				■	■					■	■
City/County - Public Works	■	■		■	■	■	■	■	■	■	■
City/County - Planning						■	■			■	
City/County - Tax Assessment						■	■				
City/County - Police/Fire						■				■	
Telephone Company	■									■	

Table 4.3. Map use.

Organization/Department	Design	Construction	Planning	Districting	Routing	Zoning	Analysis
Electric Utility	■	■			■		■
Gas Company	■	■					■
Water Company							■
City/County - Public Works	■	■			■		■
City/County - Planning			■	■		■	
City/County - Tax Assessment				■			
City/County - Police/Fire					■		■
Telephone Company	■	■			■		■

Users of Maps

Utilities such as electric companies, gas distribution companies, and telephone companies use maps for planning and designing additions, modifications, and rehabilitations to their facilities networks. Additionally, these organizations use maps in their day-to-day activities such as planning, operations, maintenance, and accounting. Municipal governments utilize maps for similar purposes, especially in instances where water and sewer facilities networks are municipally owned and operated. Municipalities also utilize maps for property assessment, land use planning, emergency response management, environmental studies, etc.

Range of Scale

For maps to be acceptable and useful to the organizations described in the preceding section, they must be produced at a scale that provides legibility and, for some applications, direct measurement capability. For land base mapping to support functions such as general land use planning, street classifications, airport influence area determinations, environmental management, etc., the maps must be relatively small scale: in the range of 1"=1,000' (1:12,000) to 1"=2 mi (1:126,720). This scale range is generally considered acceptable when map use is at a broad-brush level (i.e., the smallest potential geographic area of interest is on the order of a city block or larger). Figure 4.1 provides a sample of a typical municipal planning map.

At the other end of the scale spectrum, detailed facilities maps have historically been produced at 1"=20' (1:240) to 1"=400' (1:4,800). Such scales are appropriate when the smallest potential geographic area of interest is an individual lot or residence. Naturally, even these larger scale maps have limitations in the cases of apartment buildings, condominium complexes, etc. In those cases, standard map series are often supplemented with detailed drawings or riser (profile view) drawings to map the property or facilities information at a more suitable scale or in schematic form. Examples of utility maps are shown in figures 4.2 and 4.3.

Typical Sheet Size

The small-scale, planning-oriented sources mentioned in the preceding section are sometimes referred to as *wall maps*, because they are often mounted on a large base and displayed on a wall. Even at a scale of 1"=2,000' (1:24,000), for example, a map of a large municipality and its surrounding area of influence may measure a dozen feet across. Physical size makes it difficult to place the map in suitable storage cabinets, hence, the popular vertical wall mounting technique.

Since many of the larger scale mapping products used as data conversion sources originated in engineering environments (as opposed to cartographic environments) standard engineering sheet size nomenclature is sometimes used to describe map sizes. This is particularly true for distribution facilities maps. Thus, parties involved in map production (both pre- and post-GIS conversion) commonly mention *D-size, E-size*, etc., maps and drawings. Usually, large-scale source maps involved in GIS conversion measure between 20 and 50 inches in both dimensions. Of course, the actual mapped area within the sheet depends upon the size of the margins, position and size of the title block, etc. A fairly common mapped area within a D-size sheet, for example, is 30 horizontal inches by 20 vertical inches. Table 4.4 provides the nomenclature used for standard engineering sheet sizes.

Table 4.4. Standard sheet size nomenclature.

Size	Width	Height
A	8.5"	11"
B	17"	11"
C	22"	17"
D	34"	22"
E	44"	34"

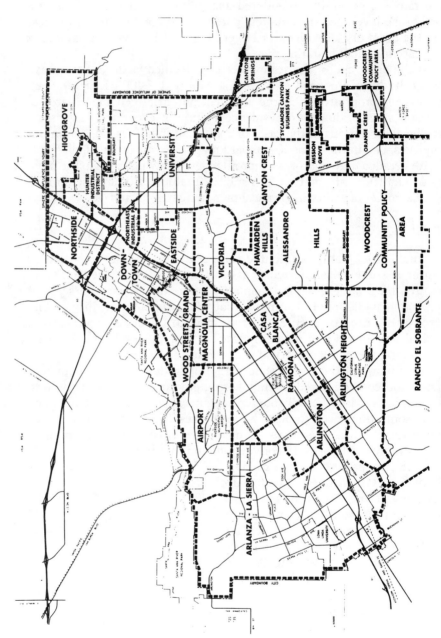

Fig. 4.1. Portion of a municipal planning map. (Source: MSE Corporation)

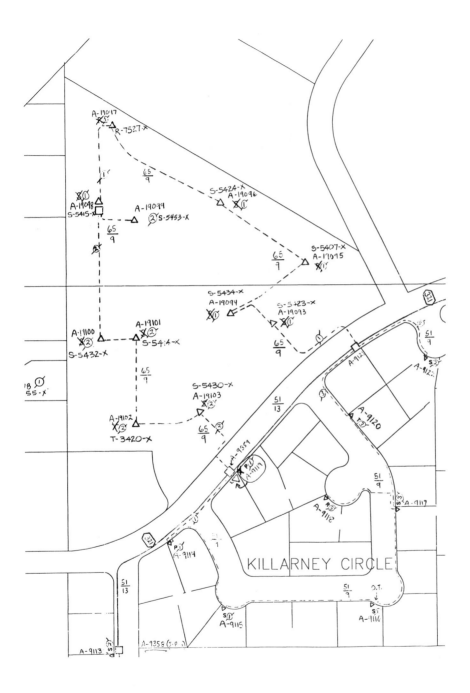

Fig. 4.2. Portion of an electric utility map. (Source: MSE Corporation)

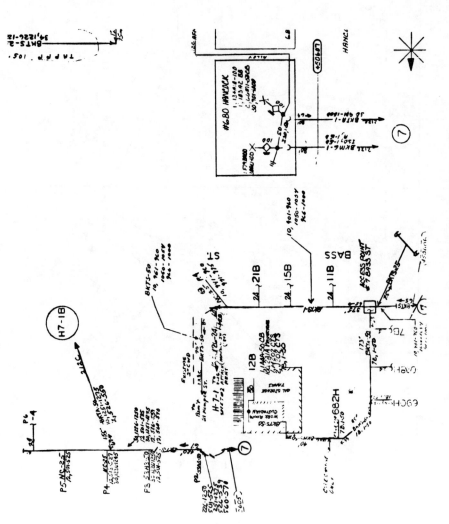

Fig. 4.3. Portion of a poorly drawn telephone facility detail.

Media

Existing source maps may have been drafted on any of the following media: mylar™, linen, paper, scribecoat, or acetate. The media of the maps actually submitted to the conversion contractor is an important consideration in the planning of a GIS data conversion project. Rarely is the client willing to provide a set of original maps to the conversion contractor. In fact, conversion contractors try to avoid the liability associated with possessing and handling a client's original documents. More typically, a paper copy (photostatic or blueline) is produced for each map involved in the data conversion process. This reproduction process is straightforward for the larger scale maps (e.g., D-size) but can present a real challenge in the case of wall maps or linen-based maps.

In some cases, a stable base material (e.g., mylar) is required for the conversion. By producing mylar copies for the conversion contractor to work from, the stretching or shrinkage problems associated with paper media can be minimized. This technique is used most frequently in cases where linear land base features require highly accurate positional replication in digital form.

Conversely, paper copies are usually acceptable for maps from which only facilities information will be converted. This is due to the fact that during conversion the facilities positions will generally be locally registered and referenced to the land base, which in most cases will have originated from a different source.

The media of the source documents is of greatest concern to the conversion contractor in those instances where the data conversion process involves digital scanning. The creases, wrinkles, stains, and background noise attendant on some types of media can diminish the effectiveness of the scanning process by significantly increasing either the scanning preparation time, postprocessing efforts, or both.

Typical Content

The smaller scale maps discussed in previous sections can contain large area facility information such as electrical primary (high-voltage) distribution lines or telephone wire center trunk lines. But usually they will be oriented toward land use, hazardous materials tracking, environmental management, and planning purposes. For example, a typical 1"=2,000' (1:24,000) map used to depict general land use might contain the following information items:

- Streets
- Street names
- Major hydrological features
- Railroads
- Polygons to depict current land use

Polygons, such as soil classifications, special redevelopment areas, bus route collection areas, and voting districts, are typically contained on separate map products. Each of these source documents contains the underlying land base (the first four items previously listed) supplemented with the particular item of interest that spawned that particular map series (e.g., zoning, annexation areas, designated historic areas, and others). Thus, an organization may have dozens of separate map products, each requiring its own data conversion strategy to ensure that the information unique to that particular map or map series is captured during the data conversion process.

Figure 4.4 provides a sample of a typical small scale (polygon-oriented) source map (wetlands).

Importance of Symbology

The reader most likely has some difficulty interpreting the maps presented in this chapter because a comprehensive symbol legend for each sample is missing. This simple fact serves to emphasize the importance of having each type of map source fully documented prior to data conversion. The conversion contractor must be able to differentiate valves from, say, manholes in order to accurately construct the digital model of a water distribution system. All source maps that require data conversion must have understandable symbologies.

In the case of small scale wall maps, a symbol legend is often included directly on the map. Moreover, the symbolization is often in the form of area shading or colorization, due to the emphasis on polygons within these types of maps.

Large-scale facilities maps may have associated symbol sets containing hundreds of different symbols. Including a comprehensive symbol legend on each sheet is therefore impractical. More often, facilities symbol appearance and usage are documented in a drafting standards manual unique to the organization. However, this is not always the case. A surprising number of organizations lack complete map production documentation. They expect personnel to memorize it. In cyclic fashion, new employees are trained by experienced personnel in the organization's practices. These situations create difficulties for conversion contractors since there is an additional effort required up front to familiarize the contractor's personnel with the project's unique symbology.

Tools Used in Original Map Production

The tools used in producing source map originals can be a major factor in a GIS data conversion project. At one end of the quality spectrum, source maps (usually on mylar) that were drafted according to a tight set

GIS Data Conversion Handbook 81

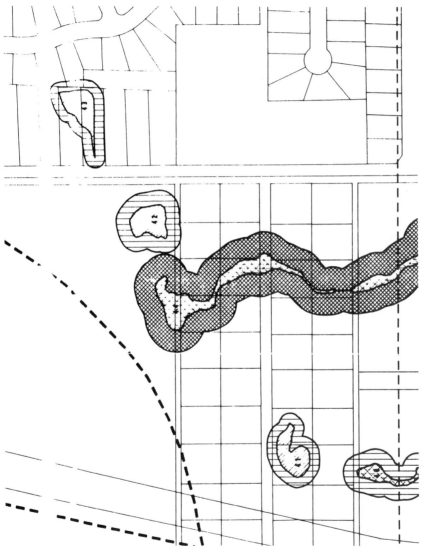

Fig. 4.4. Portion of a wetlands map. (Source: MSE Corporation)

of standards using liquid ink pens and text templates or other mechanical means of text creation, custom symbol templates, and stringent cartographic quality control procedures forbidding hand-lettered additions or postings, are straightforward to convert. Symbolization ambiguity and illegible text conditions are virtually nonexistent.

At the other end of the spectrum are pencil-drawn maps with hand-lettered text. Over time, many different lettering styles are used; the degree of legibility deteriorates, and smudges, eraser burns, and other blemishes make the source map more difficult to interpret during the conversion process.

4.4 Drawings

In the context of GIS data conversion, the primary difference between a source map and a source drawing is that the latter does not have a geographic coordinate reference system and a pervasive scale. A drawing is considered a schematic representation without geographically referenced data (e.g., a work order sketch) or a scaled drawing devoid of detailed geographic reference (e.g., an as-built drawing or a building floorplan).

Similar to the maps used by an organization, drawings to be used as sources in the data conversion process vary widely in terms of content, media, format, vintage, scale (if any), complexity, and usage. On a typical GIS project, drawings are considered to be supplemental to the source maps; maps provide an overall scaled reference frame, and the information from drawings is fitted into this frame. Drawings are used by the conversion contractor to extract information that is not shown on the maps or that the maps do not show in adequate detail. Drawings are frequently scanned, stored as raster images, and linked to a map feature. The major data conversion processes involved are the indexing and linking of the raster images.

Users of Drawings

As a general rule, drawings are used by the same organizations and departments as those who use maps. In the case of utility facilities networks, as-built drawings and plan-and-profile drawings are of precise engineering quality and often represent complex schematics of localized installations. These drawings are used by construction and maintenance crews, as well as by those involved with ordering and inventorying materials. Much of the standard construction detail in the manufacturer's drawings is omitted in the data conversion process since the GIS database contains fundamental location information and only the most important attributes for each facility element.

Accounting-oriented users rarely use drawings since these records do not contain tax district polygons and other cost or value related information. Community-oriented users such as cities, counties, and local planning organizations periodically require the use of drawings in their work. These often pertain to the construction and design of roadways, public buildings, etc. Assessment-oriented data users may occasionally use drawings in their tax assessments. This usually occurs when the valuation of a property changes due to a construction addition or renovation.

Range of Scale

Unlike maps, a collection of drawings housed at a single organization often tends to be much more heterogeneous in scale. In some cases, as-built drawings are produced by different contractors, and this results in a mixture of different drawing styles and standards. Moreover, as-built drawings are often produced at the most convenient scale, depending on the size and nature of the construction project; scales can range from 1"=10' (1:120) (for a complex civil engineering drawing) to 1"=800' (1:9,600) (for a simple cable TV installation).

An important distinction between maps and drawings is that a source map will likely contain many occurrences of a type of facility item (e.g., manholes). A drawing, on the other hand, might be a detailed view of a single manhole.

Since some drawings lack a uniform scale, they often require special processing by the conversion contractor to spatially integrate the drawing information into the corresponding GIS land base. A typical drawing will differ from a map in its content. The vast majority of information on a map is used to describe the geography. On a drawing, the majority of information deals with the design or construction of a feature such as a manhole, or a pumping station. For example, a plan-and-profile drawing for a manhole will include a small strip map for geographic orientation purposes. This plan view depicts the surrounding street(s) and, possibly, curb or sidewalk locations. To complete the drawing's original purpose, the elevation view (profile) includes detailed dimensions for the manhole, slope and fill information, material information, and a variety of construction notes.

Typical Sheet Size

Standard engineering drawing sheet sizes apply here as well. An organization usually has drawings in nearly all standard sizes (maps are generally D-size or E-size). This fact presents minor problems for the conversion contractor. The contractor's conversion procedures must accommodate a

Media

Source drawings can exist on any of the media mentioned on page 79, "Media," although a drawing is rarely created and maintained on scribecoat. Most often, drawings exist on various types of paper.

Importance of Symbology

Unlike maps, drawings rarely include a symbol legend. If drafting standards do exist for the creation and maintenance of the source drawings, they are generally in the form of a manual or handbook. The use of preprinted peel-off line types and symbols is less pronounced on drawings as compared to maps, and the use of distinct (multiple) fonts on drawings is rare.

Tools Used in Original Drawing Production

The contrast between hand-lettered penciled documents and meticulously crafted liquid ink on mylar documents is as pronounced for drawings as it is for maps. However, because many drawings are created by outside engineering, mapping, and surveying firms, a higher proportion of them are drafted using high-quality methods, including CAD.

4.5 Aerial Photographs

Aerial photographs are frequently utilized as a source for the land base component of a GIS conversion project. Photogrammetry (aerial mapping) offers an excellent way of creating a current, geographically controlled land base whose content, structure, and accuracy are precisely tailored to meet an organization's unique requirements. Aerial photography is generally done at a flight height of 2,000 feet to 10,000 feet (depending on the required level of detail); it provides the raw input necessary to support detailed planimetric and topographic mapping for a GIS implementation.

While the preferred method of acquiring aerial photographs is to solicit bids for new, custom photographs to be taken, some organizations have successfully located existing photographs that are suitable for their needs. Such photographs may have been previously produced by a federal government agency, another local jurisdiction, a local utility, or another organization with a need for geographic information.

Generally, a request for proposal (RFP) is made only after an unsuccessful search for aerial photographs has been made. Often, photographs do exist but are out-of-date or are of unsuitable scale or quality for GIS database construction purposes.

Decision to Use

The option of procuring new aerial photography is frequently taken by organizations whose existing maps are not very current and are lacking in detail and accuracy. Figures 4.5 and 4.6 show a sample aerial photograph and a corresponding aerial map.

The wealth of visual information available on an aerial photograph and on photogrammetric products usually exceeds that of the typical organization's existing maps. For example, the roof outlines of buildings and other structures provide a level of detail that is most likely not shown on the other source documents. The street locations are indicated by edges of pavement, as opposed to a stylized or schematic

Fig. 4.5. Portion of an aerial photograph. (Source: Analytical Surveys, Inc.)

Fig. 4.6. Aerial map representing photograph in figure 4.5. (Source: Analytical Surveys Inc.)

representation. Fences, billboards, and other potential obstacles for maintenance and construction crews are visible on the photographs, whereas they are most likely not shown on maps. Recent photography also offers the advantage of providing very current information as the basis for the vector information that is stereodigitized from the photographic images into the GIS database. Other information not carried on existing source documents and available through aerial photography may include sidewalks, bike paths, transmission towers, etc.

An emerging trend is to procure new aerial photography, create digital orthophotos, load the raster images into the GIS, and then vectorize only those features that are needed immediately. For example, building footprints provide valuable information for planning utility installations, but a raster image generally serves the same purpose as a manually digitized vector outline for each structure.

Typical Scales

Aerial photography can be produced at a broad range of scales. All available photographic scales are used within GIS, usually ranging from 1"=400' (1:4,800) to very small scale photographs of scanned satellite imagery. The selection of scale is influenced by a number of factors, but the most important is which features are to be visible for photointerpretation and stereocompilation. For example, if pole locations and other utility features are to be accurately located from photography during the conversion process, the photograph must have a scale of $1''=400'$ (1:4,800) or less. Photographs of a smaller scale make locating utility objects difficult.

Color vs. Black and White

A degree of controversy surrounds the choice between color or black and white aerial photography. Some argue that the increased expense associated with color film acquisition and laboratory development is not worth the resultant increase in interpretability. Hardware and software considerations related to raster display functionality and performance (larger files for color) also impact the decision in some cases.

Despite the slight cost premium, however, the trend on large municipal and utility projects is to use color products. Those who have done so are pleased with the results and are quick to point out that the extra cost for color is generally less than 1 percent of overall project costs.

Typical Features Captured through Aerial Mapping

Regardless of the type of film selected for aerial photography, a wide range of features can be captured. The process of stereodigitizing on a stereoplotter is used to capture intelligent vector data directly from the images represented on the photographs. In some cases it is easier to collect features, such as valve covers and manholes, on aerial photographs taken after *premarking* these items with paint on the street pavement. The stereoplotter operator will typically capture features such as the following, apart from topography.

- Contours and spot elevations
- Road edges and centerlines
- Sidewalks
- Buildings
- Driveways
- Fences
- Lake shorelines
- Canal outlines
- Railroad tracks
- Poles
- Manhole covers (if premarked)
- Valve access points (if premarked)
- Other premarked utility items

When digital orthophotos are utilized in the GIS implementation, the data conversion process is simplified since many of the geographic features apparent in the photo (such as rivers or paths) do not need to be discretely digitized. In addition, features that must be vectorized can be digitized using heads-up digitizing rather than the more time-consuming and expensive stereodigitizing process.

Related Intermediate Products

When utilized in a GIS data conversion project, aerial photographs are usually thought of as intermediate products. That is, they provide the basis for the stereoplotter operator to construct an accurate digital representation of the geographic features of interest. The digital vector data compiled by the operator is the end product of the process. When raster orthophoto images are used, the photographic image becomes an inherent part of the GIS land base.

4.6   Cards and Records

Many municipal governments and utility organizations maintain portions of their infrastructure and facilities information on cards or other nongraphic records. This information includes a variety of data on specific installations (i.e., a length of pipe or a manhole, street addresses, dimensions, and other descriptive details). Such cards may also include references to associated maps or drawings, information concerning the original construction contractor, maintenance records, inspection data, and special notes.

Frequently, these cards provide a textual description of the location of an item (e.g., a water main or a water service pipe) based on the nearest street intersection. Figure 4.7 provides an example.

Electric power distribution companies usually place the information collected for each pole on pole cards. The card and record data are input to the GIS database. If in card or list form, they have to be entered manually through a keyboard. Keyboard data entry can constitute a large portion of a data conversion budget. Many records exist in tabular form. Figure 4.8 provides an example of such a list.

7090 Magnolia Happy Canyon Restaurant				Serv. #37315
	Year	Type	Size	Length
17 + 10.9 Serv. off Washington	69	C	1"	27.5
				Form No. 065-41

Fig. 4.7. Sample water service card.

INVENTORY LIST		
ITEM NO.	DESCRIPTION	QTY.
70623	VALVE, CHECK, 4 IN., 600# ANSI, WEFE, RF	1
70624	ELL, WELD, 4 IN., 90 DEGREE LR, STD. W.T., SMLS, GR.-B-	2
70625	ELL, WELD, 4 IN., 45 DEGREE LR, STD. W.T., SMLS, GR.-B-	2
70626	TEE, RED., WELD, 4 IN. X 4 IN. X 2 IN., XH, SMLS, GR.-B-	1
70627	SONOTUBE FORM, 12 IN. DIA. X 48 IN, LNGTH. X 0.225 IN. W.T.	1
70628	CONCRETE, 4000 PSI COMPRESSIVE STRENGTH MINIMUM	15 CF
70629	VALVE, BALL, 4 IN., 600# ANSI, WEFE, RF, ORBIT #1462GS, BFW	1
70630	VALVE, BALL, 3 IN., 600# ANSI, WEFE, RF, ORBIT #1462GS, BFW	1
70631	VALVE, BALL, 2 IN., 600# ANSI, WEFE, RF, ORBIT #1462GS, BFW	1
70632	VALVE, PIG BALL RECEIVER, 4 IN., 600# ANSI, FE, RF, WILLIS	1
70633	VALVE, PLUG, 1/2 IN., 6000 PSI, SE, M AND F, C/W T PACKING	2
70634	VALVE, NEEDLE, 1/2 IN., 6000 PSI, SE, MARCH MOD. #N5534	1
70635	REDUCER, CONC., WELD, 4 IN. X 3IN., STD. W.T., SMLS, GR. -B-	1
70636	FLANGE, WN, 4 IN., 600# ANSI, RF, FS, MATCH 0.237 IN. W.T.	2
70637	FLANGE, WN, 3 IN., 600# ANSI, RF, FS, MATCH 0.216 IN. W.T.	1
70638	INSULATING GASKET KIT, 4 IN., 600# ANSI, PLICOSEAL TYPE E	1
70639	GASKET, RING, 4 IN., NOM. I.D. X 7-5/8 IN. O.D. X 1/16 IN.	2
70640	GASKET, RING, 3 IN., NOM. I.D. X 5-7/8 IN. O.D. X 1/16 IN.	1
70641	COUPLING, FULL, 1/2 IN., 3000# F.S., SCRD	2
70642	PIPE SUPPORT, ADJUSTABLE, 4 IN. DANIEL #1300-4	1

Fig. 4.8. Portion of a facilities record.

4.7 Existing Databases

The principal way to save time and expense on creating a GIS database is to adapt existing databases. These may already exist within the organization, or externally as developed by another GIS user for the specific project areas. Many times it is more advantageous to utilize existing databases than to rely on data conversion alone, even if the translation of the database from one GIS to another is difficult.

Most organizations approaching the implementation of a GIS will already have in place some level of computerization of geographic-related data. Depending on individual circumstances, existing digital data may be completely translated over to the GIS or may be made accessible to the GIS on an ongoing basis via a direct two-way interface. In the first case, the existing digital data is used to partially populate the GIS database, while the source system is often phased out.

In the second case, the driving design force enables both the original system and the GIS to access the data, in effect sharing the data on a continuous basis. Under either scenario, the reconciliation concerns cited previously come into play. The existing digital data must be proven accurate and reliable, especially for the specific data items that provide the linkage (i.e., the common *keys*) to the GIS.

If GIS databases exist externally to the organization, then it will be important to compare the internal GIS database requirements in detail with the external data availability and cost. In addition, assessing the practicality and cost of translating the existing data into a format suitable for the target GIS will be necessary. The database translation may be too difficult or too costly to be practical.

Community-Oriented Data

Community-oriented data often exist in digital form. These data are often available from municipalities and local government agencies, and they consist of property address files, building permit files, demographic information, zoning files, accident and crime statistics, environmental conditions, and hazardous materials information. These organizations are often able to provide these data files to others for the cost of the time spent in copying the information to a magnetic tape or tape cartridge.

Accounting-Oriented Data

Existing digital data may also be accounting-oriented information, such as a utility's continuing property record (valuation and taxation data) or client billing system. Other accounting-oriented data may include insurance records and sales information. A critical data conversion step is to reconcile client addresses in the billing system with client addresses on the maps. This will ensure that correct client loading is available to the GIS to support network analysis and load flow analysis calculations.

Assessment-Oriented Data

Assessment-oriented data are generally associated with taxation. Tax assessment departments are primarily concerned with parcel-based information. This can include utilities that are privately held and subject to local taxation. Data typically maintained in an assessment file consist of type of structure, size of structure, date built, covenants, restrictive easements, assessed value, renovations, tax history, and fair market value.

Engineering-Oriented Data

Engineering-oriented data, such as an electric utility's transformer load management data, may also exist in a database. Additionally, water and sewer system use and demand information is often available.

The objective in the data conversion planning stages is to take advantage of existing digital data to the greatest practical extent. Quality tests are as important in evaluating digital data as they are in evaluating a map or drawing as a potential conversion source.

Commercial Databases

Commercial and government databases are also potential GIS data conversion sources. However, seasoned professionals in the GIS field have learned to not rely solely on documentation describing the suitability of the digital data being considered. Often, the actual integrity and structure of the digital data differs significantly from the claims made in the documentation. The most prudent approach to assessing the suitability of existing digital data is to conduct extensive testing of the actual data.

The U.S. government offers several varieties of GIS data (land and demographic data only). The U.S. Census Bureau (USCB) offers TIGER files. These are integrated versions of the U.S. Geological Survey's (USGS) digital line graph (DLG) hydrography and transportation data and USCB's Geographic Base File/Dual Independent Map Encoding (GBF/DIME) files that were prepared for the 1990 census effort.

These types of federal data can provide a suitable low-cost solution to the problem of converting a land base in cases where positional accuracy and detailed (parcel-level) data are not requirements of the GIS. If no demographic data is required, DLG data is available directly from the USGS. An example of DLG data is shown in figure 4.9 (1:100,000 scale data).

Several private firms are refining government-provided data and, in some cases, supplementing them with additional information. The fundamental limitations of positional accuracy and the lack of detailed

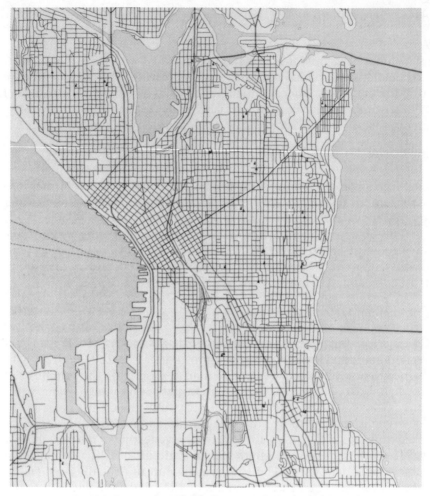

Fig. 4.9. Portion of a DLG file. (Source: U.S. Geological Survey)

topographic and cadastral information remain in many of these aftermarket products, however.

The most workable solution involving the integration of DLG or TIGER data into a GIS project seems to be a hybrid approach. For example, a large investor-owned electric utility may have a service territory covering several thousand square miles. However, only 20 percent or less of that territory may be considered urban and suburban. In such a case, it may make sense to use DLG data as the basis for the land base in rural areas and employ photogrammetric techniques to capture the land base data for the urban and suburban areas. This approach is not without problems. Edge matching linear features between urban and rural areas, for example, can be difficult. But in cases where a homogeneous solution is not an absolute requirement, it offers a cost-saving alternative.

CAD Graphic Database

Products for CAD are widely used by all engineering disciplines. Engineering firms, departments of transportation, and general contractors have been using CAD to plan, design, and draft road construction plans, architectural site plans, and property maps. The difference between these CAD drawings and GIS data is that the CAD drawings are not geographical and are based on drawing sheets as opposed to a continuous geographic land base required for GIS. Drawings with CAD are usually at a larger scale (e.g., 1"=50' [1:600] or 1"=20' [1:240]) and include more detail than is typically stored in a GIS.

Nevertheless, CAD has become an important GIS data conversion tool. Data elements such as road curblines, road centerlines, utilities, sidewalks, buildings, and property lines can be extracted from CAD files, translated into the GIS data structure, and moved to the correct geographic position in the database. Some users translate the entire drawing into the GIS, but this can significantly increase data storage requirements since the CAD drawings have many construction-related notes and text.

If an organization needs to access CAD drawings on the GIS on a continuous basis and does not require the drawings to be georeferenced for that purpose, the CAD files can be treated as GIS external files indexed to the GIS database via attribute keys. A typical example of this approach is a railroad company that constructs its GIS database from its rail construction drawings (extracted track alignment and right-of-way boundaries) and also references (through attribute database records) the digital engineering drawings of station numbers shown in the continuous track map database. This allows the railroad engineer to query and analyze the GIS database of rail facilities and to view detailed construction drawings by specifying a station number.

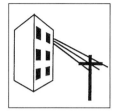

Chapter 5

Land Base Data vs. Facilities Data

5.1 Introduction

The conceptual difference between *land base data* and *facilities data* seems to be, at first glance, easy to define (see fig. 5.1).

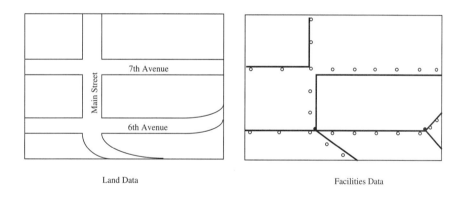

Fig. 5.1. Two-panel land/facilities data.

During the data conversion process, *raw* graphic and nongraphic data are collected and placed into a database. The database is logically structured to relate similar data types to each other, either through *layering* or *object-based* approaches. Certain types of data can be handled together, and other types of data can be seen as something separate. In many asset-oriented GIS applications, data are separated into two major groups: the land base data and the facilities/asset data. These two groups are not as easy to define as it may seem.

Probably the easiest way to differentiate between the two groups is to talk about a reference frame and objects of study that are placed into that reference frame. For example, the reference frame could be the street centerlines of a city, and the objects of study could be the pipes and valves of a water distribution system. Another example would be the parcels of a city as a reference frame and the street centerlines as objects of study. Notice that the street centerline is the reference frame in the first example and the object of study in the second.

In most cases, though, the definition is not that difficult. For example, within the GIS database of an electric utility, all the data related to facility items installed in the field are considered facility data, such as conductors and poles, and any supporting geographic information is land base. The principal concept is that, as the term *geographic information system* implies, there is a geographic reference frame against which facility data are positioned in the database. This reference frame is generally geographic in nature, showing a city, a county, or a country, but it can also be an industrial building layout, an airport, or the outline of a shopping center. Perhaps, the term *reference base* would be more appropriate.

Facilities, as the object of study, can be conductors for the electric utility, pipes for the city engineer, or buildings for the airport administration. Therefore, the difference lies more in the use of the data than in any precise definition. For purposes of discussion in this chapter, the terms *land base data* and *facilities data* will be addressed separately.

With this in mind, the next question the reader might ask is, "Why have two types of GIS data?" The answer is that land base data are often shared by many users. For example, a GIS that is installed in a city may serve, among other users, the assessor and the water department. The assessor places a zoning map over the land base, and the water engineer positions pipes over the same land base. These two users have completely different problems, and they require different solutions, but they share the land base.

Important savings are associated with sharing of digital land base data by various users, the least of which is the maintenance of only one land base for everyone to use. More and more users of land base data will promote the sharing concept and support a land base that is common to all users. Facility data, on the other hand, is not common to all but is specific to certain users. Facilities overlay the common land base that is used by all as a geographic reference.

A very important aspect of GIS data conversion is providing for feature connectivity to allow automatic tracing from one segment to the next, modeling street and facilities networks. This requires a continuous base map and total connectivity between facilities. In the database design task preceding conversion, the land base and facility data model should be established in such a way that appropriate features are related. For instance, database rules have to be established so that certain streets are interconnected and that valves, pipes, and pumps are attached to each other; this allows tracing applications, such as the modeling of traffic or water flow, to proceed through them without operator intervention.

There is usually little difference between the routing requirements along land base or facilities features. But differences remain: Traffic does not have to be *pumped* along streets, and pipes can have reverse flow, which is not the case in one-way streets.

5.2 Land Base Data

In a general sense, the term *land base data* can be defined as that part of a digital geographic database that provides a frame of reference for other information in a GIS. A land base can consist of as little as a few geographic control points to as much as a complete land database with buildings, driveways, trees, contour lines, easements, sidewalks, etc. The individual land base objects, such as street centerlines and rivers, are represented using a unique symbology for ease of capturing, retrieving, displaying, and analyzing related information in the database.

Land base objects, also called *land base elements* or *land base features*, can be subdivided into categories, grouping objects by origin and type:

-Planimetric   -Control
-Cadastral   -Administrative
-Hypsographic (topographic)

Planimetric Data

Planimetric data are mostly physical features that can be mapped from aerial photography and that have no vertical (elevation) information. In other words, these are the objects that can be seen on the ground, such as street edges, buildings, street centerlines, sidewalks, trees, etc. This data can be further subdivided into general (that is, buildings, fences, etc.), transportation (highways, airports, etc.), and hydrographic (rivers, lakes, etc.) categories. Certain planimetric data can act as facilities data, as in the case of routing along street centerlines for emergency vehicle traffic optimization.

In general, planimetric data are converted from maps by digitizing or through aerial mapping. In some cases, the planimetry can contribute to facilities conversion, as in the case of manhole and valve covers that are mapped from aerial photography or in the case of the placement of planimetry from dimensions.

Cadastral Data

All data that are related to property descriptions and ownership, such as property lines, metes-and-bounds descriptions, owner names, and addresses, are cadastral data. These data are placed into the land base by fitting them to the existing land base detail, such as aerial mapping data, or they are precisely placed with respect to a reference grid, such as a state plane coordinate system.

Some common conversion problems relating to cadastral data are address matching, street names and aliases, parcel and lot mapping, and compilation of lot dimensions. History has shown that most GIS implementations have some difficulty with matching addresses to the correct location. Problems such as duplicate addresses, no address assignment, poor standards of assigning addresses, and truncated address names have occasionally led to unforeseen costs in GIS implementations. Therefore, the conversion of cadastral data has to be based on specific studies that determine the best approach for converting addresses for the land base.

Many populous areas have more than one name for a street and have two or more streets in different locations with the same name. These difficulties come to light when implementing a GIS. Solutions to these problems may require that additional data, such as street name aliases, be placed into the database.

The data sources used for cadastral data capture are typically tax maps, but more and more users are converting legal property descriptions. The conversion of legal property descriptions is time-consuming and costly, as each property must be researched and properly *fitted* according to stringent placement standards (the Association of Property Assessors sets standards for this by state). The resulting product is a continuous cadastral fabric, created from the best possible sources, that represents properties (even if the cadastral portion of a land base is not accepted as a legal record). This product is perfectly suited to tax assessment and property mapping.

Hypsographic Data

Hypsographic data refer to elevations of the earth's surface and topographic relief. Data such as contours and spot elevations are representative examples. A terrain's hypsography is usually obtained through aerial mapping, and the products are delivered as a blanket of elevations (digital elevation model, or DEM) or as contours. In most cases, a blanket of elevations (a much less expensive alternative to contour lines) will satisfy GIS user requirements.

Control Data

To tie land base and facilities information to the correct geographic locations within the GIS database, a reference coordinate system or *grid* has to be established. In the United States, state plane coordinate systems, the Universal Transverse Mercator Grid System, or other local grids are used for GIS databases. Database information is linked to these grids with the help of objects on the ground that have precisely known coordinates. The points are called *control points*. Control points usually provide precise locations that are used to impart the appropriate shape, scale, and positioning of objects. Control data are shown on the land base map as survey monuments or geodetic control points.

Most conversion projects undertaken during the 1990s will have preparation and conversion costs associated with converting old maps referenced to the North American Datum of 1927 (NAD27) to the current standard, the North American Datum of 1983 (NAD83).

Administrative Data

Administrative data are mostly boundary data that outline an area as a unique unit for administrative purposes. For instance, this can be a municipal boundary, a tax area, a zoning district outline, a service franchise area, or a school district boundary.

Table 5.1 provides a sample list of land base data items shown by the categories previously defined.

5.3 Facilities Data

Facilities are the various pieces of equipment or other assets installed throughout a geographic area to deliver a product or service. Typical GIS facilities data are those facilities important to a user's business that are designed, installed, maintained, and removed by that user. Facilities can be individual units of property (i.e., buildings), or they can be part of a network (e.g., water distribution) that provides a particular service to the user's client base.

In a municipality the digital representation of a water network, and all the information that supports the water network serving a residential neighborhood, can be considered facilities data. In a telephone company, the facilities data could be comprised of a digital map of the buried cable network installed from the central office to all the subscriber homes, with all related information such as telephone numbers and addresses. In a gas utility, it could be comprised of all the data relating to gas distribution systems.

Table 5.1. Typical land base data.

Planimetric	Cadastral
General	• Subdivision Boundaries
	• Parcel And Lot Lines
• Building Outlines	• Rights-of-way
• Swimming Pools	• Easements
• Landfill Sites	• City Limits
• Incinerators	• County Boundaries
• Hazardous Material Sites	• Township Boundaries
• Trees/Shrubs	• Annexed Areas
• Vegetation Areas	
• Fence Lines	Control
Hydrographic	• Geodetic Control (Horizontal)
	• Geodetic Control (Vertical)
• Rivers, Creeks, Streams	• Survey Monuments
• Lakes, Ponds, Water Bodies	
• Drainage Basins	Hypsographic
• Floodplains	
	• Spot Elevations
Transportation	• Contour Lines
• Edge Of Pavement/Curblines	Administrative
• Street Centerlines	
• Street Lanes/Markings	• Map Grid Areas
• Islands	• School Districts
• Streets	• Postal Districts
• Sidewalks	• Congressional Districts
• Bikeways/Pathways	• Voting Precincts
• Driveways	• Neighborhoods
• Alleys	• Park Districts
• Railroad Tracks	• Police Districts
• Railroad Crossings	• Fire Districts
• Parking Lots/Garages	• Pest Control Districts
• Subway Lines/Stations	• Census Tracts
• Transportation Centers	• Census Blocks
• Bridges/Overpasses	• Zoning Districts
	• Historic Areas
	• Land Use Areas
	• Special Assessment Areas

Table 5.4. Sample electric facilities data.

• Poles	• Fuse Cutouts
• Towers/Structures	• Generators
• Substations	• Grounds
• Transmission Lines	• Isolating Devices
• Transformers	• Lightning Arresters
• Anchors/Guys	• Limiters
• CATV	• Manholes
• Capacitors	• Protectors
• Switches	• Padmounts
• Circuits	• Phase Marks
• Conductors	• Primary/Secondary
• Disconnects	• Regulators
• Ductbanks	• Risers
• Ducts	• Reclosers
• Fault Indicators	• Service Wire

Table 5.5. Sample gas facilities data.

• Mains	• Meters
• Valves	• Fittings
• Odorizing Stations	• Vaults
• Pump Stations	• Storage Facilities
• Services	• Leaks
• Cathodic Protection	• Drips
• Markers	• Test Points
• Regulator Stations	

5.4 Land Base vs. Facilities Base Issues

The most significant advantage of installing a GIS for a large group of users is that it allows all users to refer to the same land base. System implementation should emphasize the use of a common land base by all perceived users of the GIS. This will require additional administrative and political efforts, but the benefits to an organization will surely be noticed.

The database design should specifically reflect the applications to be supported by the system. Applications may dictate that land data be in either raster or vector format. Conversion plans should thoroughly analyze the scanning of documents and the use of scanned raster data as an alternative to traditional vector data compilation.

The GIS is usually very beneficial to those who use the system for managing their facilities, even though the cost of compiling the facilities data usually exceeds the land base data conversion costs. The accuracy of the source documents for the facilities conversion will greatly influence conversion costs. If no adequate source documents are available, implementors must use as-built drawings or costly field inventories as the data source.

Nongraphic facilities data are often converted from existing digital files. The conversion process will typically entail a system interface for remote access of data, or the bulk loading of data into the GIS database. In any case, conversion budgets should accommodate such system interface (and possibly translation) costs.

Tables 5.2 through 5.5 depict samples of facilities data for municipalities, telephone companies, electric utilities, and gas companies. These are not all-inclusive examples, but should provide the reader with an understanding of what facilities data are.

Table 5.2. Sample municipal facilities data.

• Streetlights/Signals	• Manholes
• Fire/Police Alarm Boxes	• Catch Basins
• Water Mains	• Lift Stations
• Fire Hydrants	• Flow Meters
• Metering Stations	• Treatment Plants
• Pump Stations	• Storm Sewers
• Valves	• Clean Outs
• Reducers	• Inlets
• Service Connections	• Towers/Tanks
• Sanitary Sewer Mains	• Cathodic Protection Designation

Table 5.3. Sample telephone facilities data.

• Aerial Wire	• Cable Pair Counts
• Manholes	• Capacitors
• Pedestals	• Carrier Equipment
• Poles	• Fiber Optic Equipment
• Service Drops	• Conduits
• Air Pressure Equipment	• CO Termination
• Anchors/Guys	• Cross-Connects
• Bridge Lifters	• Load Coils
• Cables (Aerial, Buried, Underground, Submarine)	• Lattice Networks
	• Serving Area Interfaces
• Fiber Optics	• Terminals
• Building Cable	• Towers/Structures
	• Microwave/Radio Equipment

The ultimate acceptance of a GIS by its users often centers on the accuracy and content of the land base data. The principal purpose (applications) of the GIS will drive the accuracy and content needs of the land base. This purpose is usually either a land-base-oriented set of applications, such as tax mapping or planning, or a facilities-based operation, such as a gas company's outside plant management.

Absolute accuracy (explained in chapter 7, "GIS Data Quality") relates to how well objects are positioned within a reference grid. This is usually a great concern in land-base-oriented applications that require more specific answers to the question, "Where exactly am I?" Relative accuracy, or the accuracy of the interval between two or more objects, is a much greater concern for most facilities data, when the question, "Where am I?" is less important than the question, "Where is the cable with respect to the curb?"

In the case of a land-base-oriented set of applications, the land-base content is usually much more complete, showing street rights-of-way, buildings, street centerlines, cadastral information such as property lines, or other detail. If the principal purpose of the GIS is to support facilities management, then the land base will usually provide only those objects that are strictly necessary as a reference framework. For example, an electric utility company may populate the land base only with rights-of-way lines, city boundaries, and rivers.

In summary, many users will need a combination of land and facilities data. Although the requirements for the land base should be defined by somebody who understands how to do this cost-effectively, the conversion of facilities data often requires extensive knowledge of that specific utility discipline to be able to develop conversion requirements properly and meaningfully.

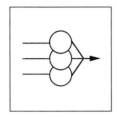

Chapter 6

Data Conversion Methods

6.1 Introduction

A variety of methods are used for converting maps, records, and existing digital data into a digital form that can be used by a GIS. Most of these methods convert information that appears on aerial photographs, existing maps, or records to a combination of vector-based images, raster images, text in a database, or text files. Other methods are used to convert a raster image of a map or record to a vector- or text-based digital file.

Usually, there are various documents and digital data to be converted, and the conversion methods used depend on the form of the original sources. The following principal forms of GIS data conversion are presented in this chapter:

- Map digitizing   -Automated conversion
- Keyboard entry   -Field survey
- Photogrammetry   -Field inventory
- Scanning   -Data translation

6.2 Map Digitizing

Map digitizing, depicted in figure 6.1, is presently the most common method of converting map data to digital data. The process usually involves three steps: data preparation, map registration, and data entry. Before digitization can occur, it is often necessary to conduct some degree of data preparation. A source map may require the addition of a new identity (map number). The conversion contractor or client may premark the map with specific points that are to be used during the registration process. Often information must be added, such as conductor segment numbers, to the maps to support subsequent data merging operations or to ensure GIS compatibility with other existing systems

105

(e.g., transformer load management). Occasionally, unnecessary information will be marked out during data preparation, although in most cases the contractor simply ignores information that is not needed. Data preparation activities may also include tracing networks from sheet to sheet, standardizing symbology, and standardizing text fields.

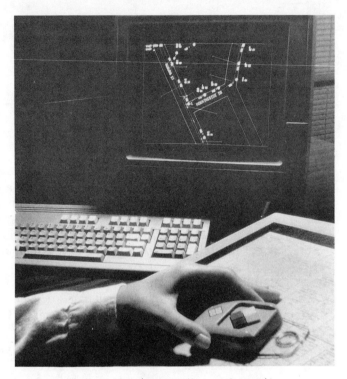

Fig. 6.1. Map digitizing. (Source: Cartotech)

Map registration is a preparatory step allowing the operator to digitize directly into a database. Map registration is a critical process, particularly in instances where a variety of sources are involved, because it will provide referencing to a common land base.

Entry of map data is commonly referred to as digitizing. Map digitizing is utilized for land base, facilities, and virtually all other GIS conversions.

Equipment

The basic hardware components necessary for map digitizing are a digitizing table, a hand-held cursor, and a computer terminal or workstation. These components are offered as packages from most computer manufacturers. Figure 6.2 illustrates a typical workstation configuration.

GIS Data Conversion Handbook 107

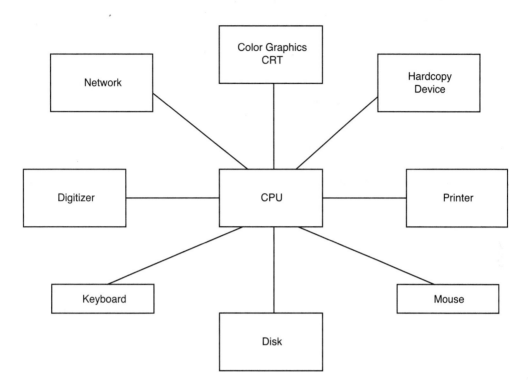

Fig. 6.2. Map digitizing workstation configuration.

Virtually all computers, from PCs to workstations to minicomputers and mainframes, can support digitizing capabilities. Workstations display the graphic features as they are being digitized. This provides the operator with a real-time check for input errors. With some older systems, alphanumeric terminals display the coordinates of a map point as they are digitized, preventing digitizing errors from being detected until a check plot is made or the data are displayed at a workstation. This is called *blind digitizing*. In modern GIS applications, the workstation platform has become the most prevalent tool, due to its sophisticated graphic display capabilities.

Setup

Figure 6.3 presents an overview of the basic steps involved in the map setup process. To digitize a map sheet into a GIS, the map must first be secured to the digitizing table surface and then registered or oriented to the database. Registration is accomplished by identifying a series of points on the map sheet and subsequently keying in their real world coordinates, or by identifying corresponding points within an existing database. For example, corresponding points shown on an aerial photograph and an existing electric distribution facilities map might be used for registration. Usually, a minimum of two, three, or four points is required for an accurate setup, depending on the mathematical solution that is implemented for the setup. After the points are entered, a coordinate transformation solution is calculated so that map coordinates can be converted to database coordinates. The results of the coordination are then displayed on the operator's screen. If the results do not meet the accuracy defined for the project, new or additional points have to be identified and measured until the setup results are in accordance with the project specifications. Printouts of map setup results that include the residual errors for each map point used can be utilized as one of the quality control tools available for conversion work.

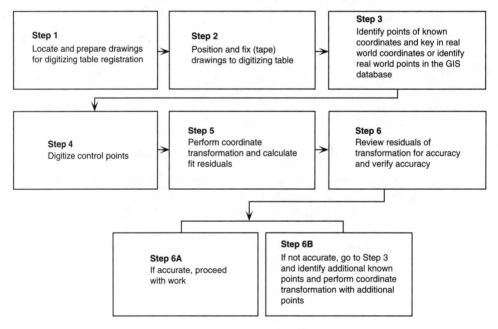

Fig. 6.3. Map setup flowchart.

Digitizing

Once the setup has been successfully completed, the digitizing process begins: the operator places the cross-hairs of the cursor over the feature to be digitized and then presses a button to relay the coordinates to the GIS. Different features can be represented by different graphic representations, such as line strings, arcs, circles, polygons, and symbols. A road, for example, would be represented by a line string while a utility pole may be a specially designed symbol and represented by a single point feature in the GIS database.

The digitizing system must contain some sort of mechanism for terminating the entry of features, such as the end of a stream. Usually, a cursor button is reserved to indicate termination of a feature.

All GIS offer some type of logical data segmentation—usually a layering or level structuring, or a grouping by feature (e.g., roads, bridges, poles, or rivers). On some systems, operators must make sure that each feature is placed on its proper layer or level. They must also make sure that the correct line style, color, or symbol is used. On other systems, this may be done automatically.

The layers, colors, line styles, and symbologies are normally determined before the digitizing process begins to help ensure data integrity. Many systems allow for the creation of customized menus that automatically set graphic element parameters and structure the data being entered according to a predefined schema.

Advantages and Disadvantages

The advantages and disadvantages of map digitizing are mostly economic in nature. The equipment is relatively inexpensive, as compared to scanners and photogrammetric instruments. Operators can usually learn a digitizing system very quickly, which limits training costs.

On the negative side is the accuracy factor. The accuracy of map digitizing is limited because it is linked to the accuracy of the original source document. Paper maps that have been used for years may have shrunk or stretched, thereby causing data entered from these documents to be less accurate than if the source were a stable base material. Digitizing large, very complex maps can be tedious, causing operator fatigue and data errors or errors of omission.

6.3 Keyboard Entry

When a digital land base is created, alphanumeric information has to be attached to the graphic objects. This is done through a computer keyboard, still the best method of entering alphanumeric information. The keyboard work involves text entry and, possibly, precision placement of additional features. Figure 6.4 depicts keyboard data entry.

Fig. 6.4. Keyboard data entry. (Source: Cartotech)

Attribute Database

One of the essential components of a GIS is its database, containing the attributes associated with the graphics data. There are some instances where the textual information for graphic entities may already exist in another database or in tabular form. In these cases, it is often possible to automatically process the textual data directly into the database. This process is commonly referred to as *bulk loading*. If no digital data exist, the attribute database must be populated by keying in the appropriate information for each field. Examples of attributes commonly attached to a property might include address, owner, owner's address, and parcel identification. Most databases allow for interactive verification of the text fields to ensure that only certain types of characters are accepted. For example, some fields may be alphanumeric, or just numeric; others may only allow a *Y* or *N* for *yes* or *no*; and so on. This helps guard against mistakes; however, as with any operator-intensive process, additional quality assurance measures must be taken to verify the data.

Code lists are also used extensively to minimize data entry. Under this scheme, a code or abbreviation is entered during conversion and the full attribute is populated later (or continually available via the code list reference). For example, a code of 01 might be used during data entry to attribute an aluminum conductor steel reinforced (ACSR). In a separate table, the code 01 will be associated with ACSR.

Graphics Database

Some textual information will generally need to be displayed with the graphics database. Most commonly, street names, addresses, water body/river names, and names of cultural features and landmarks are contained in the database either as displayable attributes or graphic text. This allows these items to be displayed on output maps and screen displays of the graphic features.

Street names and addresses can often be purchased in a digital format from value-added GIS data vendors and/or government agencies. These data can be converted to the format required by the GIS and merged with the graphic database, then cartographically edited. However, these files are rarely complete, especially for newly developed areas. Moreover, they rarely contain the full range of textual information desired in the graphics database. Therefore, much of the additional text required must be entered from the keyboard at a graphics workstation.

Several graphic parameters are associated with text. These parameters (usually entered from the keyboard) include text size, font, color, horizontal justification, vertical justification, and rotation data layer. Sometimes features need to be added to an existing graphics database, either because they are new (e.g., included in a design for an extension of a water main) or because they were missing from the source documents used for initial data entry. If the ground coordinates for these new features (e.g., poles, manholes, or hydrants) are known, they can be added by keying in the coordinate values. This technique is commonly referred to as precision placement and can apply to land features as well as to the facilities examples cited above. With GPS technology now widely available, collecting coordinates is fairly easy.

Precise Calculation

In some cases, local governments and utilities require precise positioning of water and sewer lines, gas pipes, cadastral information, etc., using coordinate geometry (COGO) type functions. To accomplish this type of data entry, the coordinates or bearing and distances of key corners or points along linear features are keyed in to locate the graphics in the database. If the other geographic data is not very accurate, fit problems between the COGO-placed features and other (possibly digitized) data may result. The better the land base accuracy, the better the fit of the COGO data.

COGO is not always needed.

Engineering Applications. Precise calculation for engineering applications involves the placement of features from utility maps or other textual information. On some utility maps, features are drawn to approximate locations, supplemented with more precise offset information. For example, a utility map may have a water main or electrical conductors shown in their approximate locations along with some text saying that the water main is offset 20 feet from the street centerline or that the buried electric conductors are offset 5 feet from the road edge. In either case, the location of the water main or electrical conductors will have to be calculated from the known coordinates of the street centerline or road edge. Utility companies also use this method for converting their right-of-way boundaries.

Sometimes the location and offset information for utility features is not kept on maps but on tabular or text records (as illustrated in chapter 4, "GIS Data Sources"). In these cases, the location of the feature must be calculated in the same manner as above.

Cadastral Applications. Cadastral or property boundary information has become a very important layer of information in many GIS. Parcel and property records can be linked and associated with a wide variety of features and attributes ranging from deeds and ownership records to tax assessor records and utility services. Versatility has made the cadastral layer one of the most basic components of a GIS.

Many cadastral records are kept on outdated or local coordinate systems. By converting the cadastral records to a GIS, the records become referenced to a common coordinate system, such as state plane, with the land base and facilities. To enter all the property corners precisely, COGO functions are employed, using angle and distance data to calculate property corners from key cadastral points such as section corners. Although this is the most accurate method for converting cadastral data, it is also expensive. An alternative approach is to precisely enter only the key cadastral corners and perform a best-fit adjustment for the individual parcels. This method is far more economical and still provides a solid spatial solution for the property boundaries.

6.4 Photogrammetry

Photogrammetry is defined as the art, science, and technology of obtaining reliable information about physical objects and the environment through processes of recording, measuring, and interpreting photographic images. Today, there are two distinct areas of photogrammetry: interpretive and metric. Interpretive photogrammetry includes photographic interpretation and remote sensing, and deals with identifying objects and determining their significance. Metric photogrammetry involves making precise measurements from photographs and other images to determine the specific location of objects. Aerial metric photogrammetry is most often used for the preparation of planimetric and topographic maps.

Although the primary function of photogrammetry in GIS is the creation of a land base, it can also be used to convert facilities information visible on the photographs. Features such as transmission towers, utility poles, and streetlights are easily identifiable on most scales of photographs used for land base conversion. With large-scale photography (that flown at lower altitudes), facilities such as manholes, culverts, and fire hydrants become visible. In most cases, these facilities are premarked on the ground with highly visible paint before the photographs are taken.

An interesting evolution is occurring in this area, however. Because of the rapid expansion of GPS technology, premarking facilities is giving way to a process that is a hybrid of surveying and field verification.

Proponents of this new technique state that it is more complete, can be accomplished at approximately the same cost as premarking, and has less environmental impact.

Interpretive photogrammetry also has some function in GIS data conversion. An example is a natural resource forest inventory GIS. An experienced forester with photogrammetric skills can accurately identify forest types, tree heights, and stand densities from aerial photographs. These features can be identified and digitized using stereoscopic digitizing equipment, or the photographs can be scanned and the interpretation done using image analysis software.

Photogrammetry takes advantage of two basic principles: the geometry of vertical photographs and the human ability to see stereoscopically. Vertical photographs are those exposed with the axis of the camera oriented vertical (as much as possible) to the earth. In photogrammetry, objects are viewed stereoscopically from the perspectives of two different photographs. The left eye views one photograph while the right eye views the other. This is the same principle used by three-dimensional movies.

Photogrammetry Process Overview

Several tasks must be accomplished in order to effectively use photogrammetry to create a GIS land base. Figure 6.5 presents an overview of the photogrammetry process. The first task is to establish ground control points. These are points whose ground coordinates have been determined by surveying and then marked with highly reflective material so that they can be seen on the photographs. Once ground control points are established, the flight mission can take place. Photographs are taken in strips with enough overlap between photos and side lap between strips to ensure stereoscopic coverage of the entire area.

After the photographic mission, aerial triangulation is performed. This is the process of adding ground control points so that there are enough available for the setup of each stereoscopic model. Finally, pairs of photographs (*stereomodels*) are placed in instruments known as *photogrammetric stereoplotters* and subjected to a process of orientation and measurement. When this is complete, features are digitized.

Why Use Photogrammetry?

The most compelling reason to use photogrammetry to create a GIS land base is the high degree of accuracy it provides. National Map Accuracy Standards state that 90 percent of principal planimetric features are required to be within 1/30 inch of true positions for map scales of 1:20,000 or larger and 1/50 inch for map scales smaller than 1:20,000.

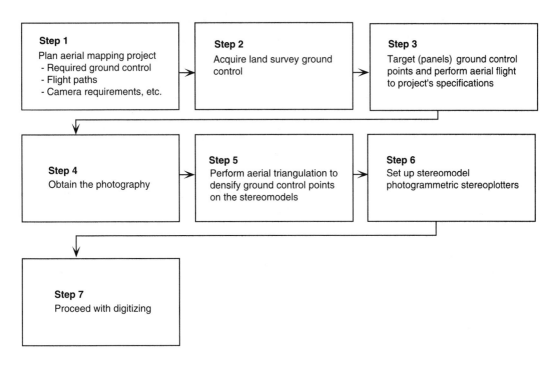

Fig. 6.5. Photogrammetry process.

On a USGS quad, for example, the scale is 1:24,000 (1"=2,000'); therefore, the allowable map error is 40 feet (1/50 inch x 2,000 feet). For elevations, the standards require that 90 percent of all points tested be within half the contour interval. These accuracies are easily obtainable using photogrammetry. They are absolute accuracies; that is, the allowable errors are based on an object's true location. Relative accuracy, the distance between two points or objects on the ground versus the distance in ground units of the same two points on a map, may be required for some GIS applications. For instance, it may be more important to an electrical utility to know the relative distance between transmission towers than to know their absolute positions. Chapter 7, "GIS Data Quality," contains detailed information pertaining to accuracy.

Most utility company maps, assessors maps, and others do not meet National Map Accuracy Standards; some are not even close. Therefore, the real advantage to using photogrammetry is the resultant accurate land base to which facilities information can be referenced and placed or to which cadastral or other types of information can be fit. However, this process involves fitting the less accurate maps to the more accurate, photogrammetrically compiled base. This can present difficulty, especially in instances where the end users have perpetually overestimated the accuracy of the existing map set(s) prior to conversion.

Ground Control

To make precise measurements from aerial photographs, known accurate control points must be identifiable on the photographs. This is accomplished by placing panels on surveyed points that are large enough to be seen on a photograph at a given scale. There are three types of control points: horizontal, vertical, and full. Horizontal control points contain easting and northing coordinates; vertical control points contain elevations; and full control points contain both. In North America, horizontal points are usually based on NAD83. In the United States, the USGS has developed state plane coordinate systems based on NAD83 so that each state has a continuous local coordinate system. Vertical points in North America are based on the 1929 North American Vertical Datum of the National Geodetic Survey (NGS29).

The amount and type of control used on any particular GIS project depends on several factors, such as the desired accuracy and the scale of photography. If a GIS does not require contours or a digital terrain model, the amount of vertical control can be greatly reduced. Obviously, the more control points placed, the higher the cost. Typically, two horizontal control points are placed at both ends of each aerial photography strip plus another control point about every third or fourth stereomodel. Usually, there are at least three vertical control points per photograph.

Photography

The acquisition of photography is a very important step in photogrammetry. As mentioned earlier, the photography must provide stereoscopic coverage of the entire project area. The standard format size for aerial photographs is 9 inches x 9 inches. Photographs along a strip overlap (in the forward motion of flight paths) by about 60 percent. This overlap is known as *endlap* and ensures stereoscopic coverage of all points along the strip. The endlap area from the center of one photograph to the center of the next photograph is known as the stereomodel. The overlap between strips of photographs is about 30 percent and is known as *sidelap*. Sidelaps of 30 percent mean the top and bottom edges of the photographs do not have to be used for the stereomodel or aerial triangulation. Since this is the area where distortions are greatest, these edges should be avoided. Displacement of features due to ground relief increases with distance from the nadir of the photograph.

Aerial Triangulation

Aerial triangulation is a complex mathematical process that allows the formation of a photogrammetric block of stereomodels. During the process, all the stereomodels covering a project area are linked into a continuous blanket of coordinated points, allowing sparse control points to completely define the position of each stereomodel for setup. At the same time, this process allows the detection of faulty control points prior to data collection. The purpose of triangulation is to supplement the ground surveyed points with additional control points that can be used to strengthen the solution. The *calculated* control points resulting from triangulation are much less costly (on a per point basis) than control points requiring field work.

Stereoplotters

Stereoplotters are used to view two consecutive aerial photographs in three dimensions. Stereoplotters have three basic systems:

- A projection system for creating three-dimensional stereomodels,
- A viewing system that allows the stereomodel to be observed, and
- A measuring system that can make and record measurements from the stereomodel.

There are two types of stereoplotters: analog and analytical. Analog stereoplotters are mechanical-optical devices on which settings are performed by mechanical adjustments. This type of stereoplotter has given way to the more accurate and dynamic analytical type of stereoplotter, which is becoming the industry standard. Analytic stereoplotters differ from analog stereoplotters in that they are computer driven.

Digitizing

Stereodigitizing can be done from either analog or analytic plotters, if properly equipped. Once a stereomodel has been successfully set up, the features and elevations of the model can be digitized. At this point of the photogrammetric process, there are many similarities to map digitizing. The main difference is that instead of a digitizing table, cursor, and source map, there are now a stereoplotter, a measuring mark, and a three-dimensional photographic image.

Planimetric data visible in a stereomodel are digitized by placing a measuring mark on them and by pressing a button (hand or foot controlled). Building outlines, street centerlines, lakes, and other planimetric detail are then captured.

Elevations can be collected from aerial photographs using two basic methods: contouring and digital terrain models (DTM). Contouring is a technique where the measuring mark is set to a constant elevation value and then moved along the ground at that elevation in the stereomodel. At the same time the measuring mark is moving, a line string is created to indicate a contour line of that elevation. Then the elevation of the measuring mark is increased by the amount of the contour interval and moved across the model again. This process is repeated by the stereoplotter operator until all the contours for the model have been completed.

Digital terrain models are collected by systematically digitizing a series of spot heights throughout the model and digitizing breaklines, which represent breaks in the terrain that are both natural (such as ridges and drains) and artificial (such as road edges and retaining walls). The breaklines and spot heights are combined and formed into a triangle network. From this triangle network, contours can be generated. In addition to contours, DTM can provide slope and aspect determination, drainage and flood analysis, orthophotography, road alignments, and cut and fill calculations. Digital terrain models offer much more to GIS than contouring alone and are becoming widely used.

The preceding paragraphs present an extremely simplified view of a very complicated process. Qualified stereoinstrument operators are difficult to find and the training required is intensive. The process itself is labor intensive, and operator experience is vital in creating quality results. For these reasons, coupled with the high cost of stereoinstruments, the cost of topographic data entry is high compared to other processes such as scanning or manual digitizing.

Orthophotography

Orthophotographs, or orthophotos, are a photogrammetric product useful in GIS applications because of the wealth of visual information they provide. Much of this information may not be included in the GIS database (e.g., vegetation, fences, or billboards), yet will be useful to the client for planning, engineering, and operations purposes. An orthophoto shows images of objects in their true position, much like a line map. With a photograph, objects away from the center appear to fall away from it. For example, a straight highway running over a hill will look bent on the photograph. This is known as *relief displacement*, and is true for ground surfaces as well as objects. In an orthophoto, the road will appear as it really is: straight. Orthophotos are produced by taking elevation information, such as a DTM, and combining it with an original photograph using an optical projection instrument that actually moves small portions of the photographic image laterally to make an orthonegative. The orthonegative can then be processed photographically.

Orthophotos have the advantages of both aerial photographs and line maps. They contain all the imagery of aerial photographs so all details that are visible become part of the map. Usually, there are far more objects on a photograph than can be placed on a map. Since orthophotos are geometrically correct, it is possible to make accurate distance measurements. Vector and text information can also be added to orthophotos. This is useful for cultural information, such as street names, and for elevation data, like contours and spot heights.

Future Advancements

Three new advances in photogrammetry will have an impact on GIS. They are: digital orthophotography, soft-copy photogrammetry, and GPS-controlled image capture.

Digital orthophotography, or *ortho images*, uses digital processes that remove relief displacement by moving image portions digitally instead of photographically, after photographs are scanned with high resolution scanners (up to 4,000 dpi—the resolution of aerial photography is about 2,400 dpi).

Soft-copy photogrammetry is based on the use of data on scanned images of photographic stereomodels appearing on a pair of cathode-ray tubes (CRT) that are mounted in the stereoplotters. Finally, use of GPS-controlled image capture promises to greatly reduce the amount of ground control work now conducted by field survey. This technology involves attaching GPS equipment to a precision aerial camera platform.

6.5 Scanning

Scanning is the process of converting paper maps and documents into a digital raster format. Scanners come in a variety of types and styles and are becoming more commonplace in data conversion for GIS. There are two basic styles of scanners: flatbed and rotating drum. In flatbed scanners, the document is placed on a flat glass surface while the scanner's sensing device moves across the image. In some flatbed scanners, such as that shown in figure 6.6, the sensing device remains stationary while the document moves across the sensor. Rotating drum scanners mount the document on a rotating drum so that the document becomes a part of the drum's circumference. The light source is located in the center of the drum. Many rotating drum scanners are plotters as well as scanners.

Scanners have two basic types of operation: reflective and transmissive. Reflective scanners have the light source and sensor on the same side of the document. This type of scanner is necessary for opaque maps and documents. Transmissive scanners shine the light through

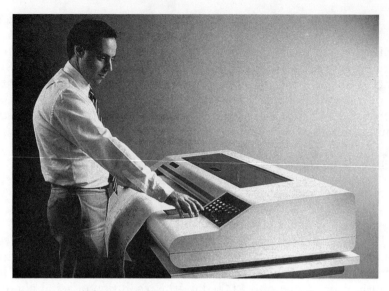

Fig. 6.6. Scanning operations. (Source: Smartscan)

the document to the sensor which is located on the other side. Transmissive scanners generally produce higher quality images since there is less light scattering. These scanners are used for transparent documents such as mylar map separates and photographic diapositives.

Digital data from scanners are in a raster format, or a row-and-column grid of picture elements or pixels, each with a numeric value. Pixel values can be as simple as on/off binary values; however, they usually are recorded as an eight-bit value from 0 to 255. The eight-bit number is often referred to as a pixel's brightness value. Color images are formed by scanning a document three separate times with red, blue, and green filters, creating three separate files with pixel values of 0 to 255.

The resolution of a raster image can be measured in three different ways: dots (pixels) per inch (dpi), actual pixel size, and the object pixel size (often used for satellite imagery). *Dots per inch* actually means pixels per inch. Therefore, if a file is 400 dpi, it has 400 pixels per inch in both the x and y directions. Actual pixel size refers to the dimensions of the pixel itself, and is usually expressed in microns, such as 25 microns. A pixel size of 25 microns translates to 1,016 dpi and 400 dpi translates to 63.5 micron pixels. Raster satellite imagery resolution is expressed in the ground area each pixel covers. Spot panchromatic imagery, for example, has a resolution of 10 meters x 10 meters. Since each pixel usually requires eight bits of memory, the relationship between file size and resolution becomes obvious. If the resolution of a file doubles from 400 dpi to 800 dpi, the number of pixels will quadruple, as will the file size (before compression).

Scanning a Document

The majority of scanned documents for GIS data conversion are facility data or land fabric cadastral data. These documents are almost always paper; therefore, reflective scanners must be used. Once the scanning procedure has taken place, the scanned image can be brought up on a graphics screen. Even though the document is now in a digital format (raster), it still needs to be converted to a vector format. This can be accomplished either by geographically registering the image and digitizing it—a process similar to map digitizing but termed *heads-up* digitizing—or by using conversion techniques that automatically trace raster lines and vectorize them.

One big advantage of scanning facility maps or other related paper maps is that, once they are all in a raster format, they can all be made to the same scale. For example, if a water facilities map is 1"=50' (1:600) scale and a storm sewer map is 1"=100' (1:1,200) scale, they can now be directly overlaid on a graphics workstation and compared. However, critics of scanning point out that the ability to warp and register various documents in raster format depends on the individual software package and is the most uncertain/unproven area of scanning.

Many documents that require scanning for GIS data conversion are either old or in bad condition. When these documents are scanned, there is often a lot of noise or speckle in the raster image. In these cases, it is usually necessary to perform some image processing functions to remove the speckle and noise from the raster image before proceeding with vectorization.

There is also a growing need to manage text documents that are related to GIS. By scanning these documents, they can be stored in a raster format and integrated with a GIS. This allows users to access information about individual parcels or facilities that may not be stored in the GIS attribute database.

Image Storage

Raster files tend to be exceedingly large. A 36-inch-x-24-inch map sheet that is scanned at 400 dpi can take up to 138 megabytes of memory if each pixel is eight bits. Fortunately, most raster files can be compressed. Typical compression formats are CCITT Group IV (international compression standard) and tag image file format (TIFF).

Another common format is Run Length Encoded (RLE). In an RLE file, all sequential pixels with the same brightness value are stored as a brightness value followed by the number of pixels containing this value. This format is particularly useful for scanned map sheets with line work. However, it is less effective for scanned images such as aerial photographs and satellite images since they normally have a more heterogeneous structure of brightness values.

The technological advancement of optical disks has also helped alleviate the storage problems of raster files. Optical disks are considered to be a more efficient medium for storing scanned data than conventional magnetic disks and magnetic tapes. Three types of optical disks are used today for image storage: the CD-ROM, the WORM, and the erasable disk. Compact-disk, read-only memory (CD-ROM) can only be read. Write-Once, Read-Many (WORM) disks are more useful since the user can control the input and then distribute them to be read. Erasable disks represent the latest technology in optical disks; these disks can be written to, read, and erased as often as the user wants. Erasable disks provide the flexibility of current magnetic technology with the increased storage capability of optical disks.

The drawbacks to optical disks are their cost and access speed. The cost of optical disk storage is about $1 per megabyte on an average system. If the access speed of the image data is important, then the data that has to be readily available is transferred to magnetic disk first.

Uses for Scanned Images

As scanning technology continues to improve, the uses for scanned images continue to grow, especially for data conversion. Scanning facility and property maps for vectorization will be a major use, and will continue to increase as automated conversion techniques continue to advance. The scanning of irreplaceable documents allows the safe handling of raster images instead of the documents. Scanned aerial photographs have several uses for GIS data conversion. They can be warped to ground control or orthorectified to produce digital orthophotos. Both of these applications can be used for heads-up digitizing of land base features. Digital orthophotos can be used as an actual layer of data in a GIS as can raster satellite imagery. Scanned aerial photographs will also be an integral part of soft copy or workstation photogrammetry. Another conversion application is quality control of stereodigitized data. The raster image of the photo can be overlaid with the stereodigitized vector data and checked for completeness and quality.

Scanning nonmap or text documents, known as document imaging or electronic document management (EDM), is also beginning to play an important role in GIS. Electronic document management systems support efficient and safe document storage in a single location accessible to several users. This eliminates the need to copy and store documents in several places. Image documents can be linked to the specific graphic objects they refer to in a GIS for retrieval by keys stored as attributes.

6.6 Automated Conversion

Automated data conversion for GIS consists of creating vector graphics and associated database attributes from raster data. Automated conversion systems require specialized hardware and software, and usually include scanners, image handling, and recognition processing. The largest cost associated with a GIS is data conversion; automated conversion offers the potential to reduce this cost. Another potential benefit of automated conversion is the capacity to handle larger volumes of source maps and drawings, thus reducing the time it takes to complete the conversion. This is significant since large data conversion projects can take several years to complete. Digitizing large volumes of map sheets can be a tedious and fatiguing task, which is bound to lead to errors in the manual conversion process. Automated conversion offers the potential of higher accuracy and quality by using automated validation and overlay with source images. The basic automated conversion process is as follows:

- Scan source maps to create raster files.
- Recognize lines, symbols, and characters automatically.
- Construct vector elements and text.
- Form related objects.
- Insert new data into an intelligent database.

Automated conversion is a new technology that has not yet replaced manual conversion as the preferred and most practical method of data conversion. However, as the demand for GIS continues to grow, advancements in automated conversion technology are sure to continue. Currently, most conversion contractors agree that the most effective method of conversion is not fully automated, but rather, interactive (operator-assisted) vectorization.

Lines and Curves

Map line work is the simplest of map data to recognize and vectorize. This area of automated conversion is by far the most advanced. Line work in a raster image is a connected group of pixels. These are recognized by the software and converted to lines, arcs, or curves. These new vector elements generally require further processing to produce aesthetically pleasing graphics and, for most GIS users, to support connectivity. Vectorized lines often appear to be rough and jagged. This can be caused from the resolution of the raster image and the vectorization algorithms. The lines usually need to be lengthened so that they are more representative of the feature being depicted, such as a road segment or water main. This lengthening process also serves to make the lines more aesthetically acceptable. In some cases, a series of short vectors may be replaced by an arc or a curve. Cleanup functions such as line snapping and the elimination of overshoots and undershoots are also required.

Maps frequently use various dashed line styles to represent different facilities and features. Recognition of these patterns is still an area that is not fully functional and requires further development.

Symbols

Utility maps are often characterized by a large variety of symbols, line styles, and extensive textual information. Some utilities even use different symbols to represent the same feature depending upon which set of maps is being used. This complexity requires that each set of maps have its own set of rules. Symbol recognition is not often used in automated conversion; further development of automated rule base generation holds the most promise for future use of symbol recognition.

Text

Systems frequently use optical character recognition (OCR) techniques to recognize text characters. Generally speaking, text recognition is further developed than symbol recognition; however, the results of these techniques vary widely in generality, performance, and accuracy. There has been little success with the handwritten or rotated text often contained on utility maps.

Batch vs. On-Line Conversion

Originally, automated conversion was a batch process performed on the raster file. This forced the conversion software to try to make intelligent decisions on its own. In cases where lines cross or touch, or gaps occur, software has been very limited in its ability to make an intelligent choice about how to proceed. This situation often resulted in the need for extensive post vectorization editing; thus, manual digitizing was frequently more efficient than automatic vectorization.

Recently, there have been advancements in on-line or semiautomatic conversion. Here, the same principles are used as in automatic conversion, but in an interactive graphics mode. This allows an operator to make the intelligent decisions that automated software had trouble making. When the on-line system comes to a point where lines intersect or gaps occur, the operator directs the software how to continue with the vectorization process.

6.7 Field Survey

The purpose of surveying for a GIS is to locate the position of points on the earth's surface. Field surveys provide several services for GIS; the most common is establishing photo control for analytical aerial triangulation. Other important services are locating coordinates for new features, determining cadastral corners, and verifying land base data. Figure 6.7 depicts conventional field surveying.

Fig. 6.7. Conventional field survey. (Source: MSE Corporation)

Conventional Field Survey

There are basically two types of field surveys, horizontal and vertical. Horizontal surveys determine a point's X and Y (easting and northing) coordinates on the earth's surface. Most U.S.-based GIS projects use state plane coordinate systems established by the USGS. International-based GIS often use Universal Transverse Mercator (UTM) or other local coordinate systems. Vertical surveys capture Z coordinates (elevation data); these are usually based on the mean sea level.

Horizontal surveys can be conducted using various combinations of measuring horizontal angles and distances. Four methods are prevalent: GPS surveys, trilateration, triangulation, and traversing. With GPS surveys, coordinates of points on earth are calculated from distances measured to satellites. Trilateration is the measurement of distances between points forming horizontal triangles. Triangulation (the measurement of horizontal angles) is rarely used anymore, except to check trilateration results. Traversing is the linking of a series of horizontal distances that are held into position by measured angles. Angles are still measured with theodolites, but distances are now measured electronically.

Vertical surveys are done in three main ways: GPS surveys, differential leveling, and trigonometric leveling. A GPS survey is able to provide elevations as well as horizontal positions, and it is already being used extensively to create large-area DEM data for three-dimensional surface generation. Differential (or spirit) leveling is still considered the most precise type of vertical survey, as it provides precision to fractions of an inch, but it is also the most expensive method. Differential leveling is a painstaking method that can be observed along highways where alternate readings are taken on vertical graduated rods. Trigonometric leveling uses theodolites and electronic-distance measuring devices to observe angles and long distances between points to obtain positions and elevations. This method is generally much less accurate than spirit leveling, but is also much less costly.

Another method that is used for *control densification*, or the creation of ground points with coordinates, is photogrammetry. Generally, the accuracies of photogrammetric surveys parallel trigonometric leveling results.

There are different levels of accuracy standards for both horizontal and vertical surveys established by the Federal Geodetic Control Committee (FGCC). The levels are known as order and class and are shown in table 6.1.

The degree of accuracy used is determined by the accuracy requirement of a particular GIS. Typically, second-order horizontal control is used for purposes such as photo control, and first-order horizontal control is used for setting permanent control monuments that can be used to establish future surveys in the area. For vertical surveys in GIS, first-order accuracy is rarely used due to its high cost. Second- and third-order accuracies are more commonly used in these surveys.

GPS Surveys

The Global Positioning System (GPS) represents a revolutionary new technology for field surveying. A GPS is a three-dimensional surveying system based on radio signals from the NAVSTAR constellation of

Table 6.1. Horizontal and vertical accuracy levels.

Classification	Limiting Error Level	Recommended Uses
First Order	1 part in 100,000	Primary National Network, Metropolitan area surveys, scientific studies
Second Order Class I	1 part in 50,000	Area control that strengthens the National Network; subsidiary metropolitan control
Second Order Class II	1 part in 20,000	Area control that contributes to, but is supplemental to, the National Network
Third Order Class I	1 part in 10,000	General control surveys referenced to the National Network; local control surveys
Third Order Class II	1 part in 5,000	General control surveys referenced to the National Network; local control surveys

Classification	Limiting Standard Error	Recommended Uses
First Order Class I	$0.5 \text{ mm} \sqrt{K}$	Basic framework of the National Network and metropolitan area control; regional crustal movement studies; extensive engineering projects; support for subsidiary surveys
First Order Class II	$0.7 \text{ mm} \sqrt{K}$	Basic framework of the National Network and metropolitan area control; regional crustal movement studies; extensive engineering projects; support for subsidiary surveys
Second Order Class I	$1.5 \text{ mm} \sqrt{K}$	Secondary framework of the National Network and metropolitan area control; local crustal movement studies, large engineering projects; tidal boundary reference; support for lower order surveys
Second Order Class II	$1.3 \text{ mm} \sqrt{K}$	Densification within the National Network; rapid subsidence studies; local engineering projects; topographic mapping
Third Order Class I	$2.0 \text{ mm} \sqrt{K}$	Small-scale topographic mapping; establishing gradients in mountainous areas; small engineering projects; may or may not be adjusted to the National Network

K is the distance in kilometers between points

earth-orbiting satellites. The system has been implemented by the U.S. Department of Defense (DOD) and provides a worldwide navigation system. More than 24 satellites will comprise the NAVSTAR constellation (see fig. 6.8).

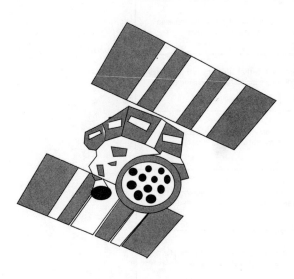

Fig. 6.8. GPS satellite.

Global Positioning System receivers, which can be small handheld electronic devices, are used to receive radio signals from the satellites. To be effective, at least three satellites need to be visible for horizontal surveys and four satellites for vertical surveys. The GPS receivers process the satellite signals and give positions based on the World Geodetic System 1984 (WGS-84), which is an earth-centered, earth-fixed coordinate system. These WGS-84 coordinates can then be translated into whatever coordinate system a particular GIS requires. The GPS receivers have a great range of capabilities; they vary in the number of channels for reading signals (the more channels, the better).

A drawback to GPS is selective availability. Since the service is provided by the federal government and used by the DOD, defense information emitted by satellites is modified to reduce the satellite's usefulness. At times, this modification is removed, allowing much greater accuracies. Fortunately, methods have been developed to allow accurate positioning, especially with *differential positioning* where two receiver stations work in tandem.

The GPS is quickly becoming the standard for surveying, and first-order accuracies are achievable using these differential positioning techniques. The results obtained from GPS can also be used to establish baseline vector components between stations.

6.8 Field Inventory

Field inventory often includes identifying features such as facilities; recording the geographic location of these features; and recording the connectivity, size, material, etc., of the features. Graphics and alphanumeric data are provided as input to the GIS database.

Data Collection

Data collection by field inventory methods requires physical visits to the site of the feature or utility that needs to be entered into the GIS database to examine and record actual conditions. Figure 6.9 depicts field inventory operations. The extent to which this method is used in GIS projects varies from not at all to the undertaking of complete inventories. Field inventory data collection is typically used as a supplement to other data conversion methods and not as a primary means of data acquisition. Field inventory can supplement photogrammetry when features such as roads, buildings, or utilities have been built after the acquisition of the aerial photography or when features are obscured on the photographs due to heavy tree canopies, snow, or clouds. Field inventory can supplement map digitizing when the maps are out-of-date or incomplete.

For organizations with existing facilities or cadastral records of questionable quality and currentness, a complete physical field inventory is the very best means of assuring that the converted GIS database will accurately reflect real-world conditions.

Others utilize field-portable computers, using customized database forms, for the entry of attribute information pertaining to a feature. With this approach, the graphical drawing of the feature is then done on a map. The computer database can be used to verify and edit the attribute data while in the field. The attribute files are later transferred to the GIS database, and the graphic information is digitized from the maps.

Data can be captured during a field inventory in a variety of ways. Some projects utilize paper maps and sketches to record coordinates, measurements, attributes, and features. This information can later be digitized or entered into the GIS database. Some use audio or video

Fig. 6.9. Field inventory operations. (Source: Cartotech)

recording devices to capture information. Such information is then processed for input to the GIS.

Graphics capability on portable computers can eliminate the need to carry both the computer and a set of maps into the field. The graphics capability could provide editing functions so that no further processing would be needed. Laptop, notebook, and hand-held computers are becoming more powerful and more capable of a wide range of applications, many of which have potential for field inventory work.

Data Verification

Data verification by field inventory is a common way to perform quality assurance checks for positional accuracy and completeness. Field crews usually bring measuring devices, such as survey equipment, to test absolute and relative accuracies of the GIS land base. Facilities data verification is also completed using field inventory techniques. Crews often validate facilities information on a periodic basis to record and monitor damages, obstructions, and other conditions affecting the equipment.

In general, a variety of data conversion methods exist, and it is the data requirements of a project and the creativity of a conversion contractor that influence the selection of the specific conversion methods to be used. Therefore, a project's requirements should be carefully analyzed and planned before any data conversion begins. Chapter 12, "Developing a GIS Data Conversion Plan," provides additional detail concerning the proper sequence of steps leading up to a full conversion effort.

6.9 Data Translation

Data translation software is often used during the initial steps of the conversion process to either (1) translate data from a stereocompilation system to the conversion system that will be used to convert the facilities data, or (2) translate existing digital data (e.g., TIGER files or data from another entity) into the conversion system format. On the output end of the process, it is common for the conversion contractor to translate from the proprietary system used during the actual conversion process to a format that matches the client's destination GIS.

Chapter 7

GIS Data Quality

7.1 Introduction

Data quality is a significant requirement in any GIS data conversion project, and it can have a major impact on data conversion method selection and cost. The required data quality for an individual GIS project is dependent on that project's anticipated applications and output products. Ultimately, the quality of the database will contribute to the confidence the users of the GIS will have in its data.

There are three categories of data within GIS for which data quality has to be defined: the graphic features, with their position and representation; the attributes; and the database intelligence. For these three categories, data quality can be thought of as how closely the data represented in the GIS database match the information in the real world. The categories have similar data quality components, such as completeness, correctness, timeliness, and coverage, but the graphic features are also subject to cartographic quality considerations. Therefore, data quality discussions in this publication are split into (1) cartographic quality, which deals with graphic accuracy and graphic quality, and (2) informational quality, which deals with correctness, completeness, timeliness, and integrity of graphic data, attributes, and topology. Cartographic quality is discussed first, followed by informational quality.

7.2 Cartographic Quality

Three types of cartographic quality govern the graphic representation of objects; they are relative accuracy, absolute accuracy, and graphic quality. The first is a measure of how objects are positioned relative to each other; the second indicates how closely objects are positioned within an absolute reference frame, such as a state plane coordinate system;

and the third refers to graphic representations of the digital maps that are correct and legible.

Relative Accuracy

Relative accuracy is a measure of the maximum deviation between the interval between two objects on a map and the corresponding interval between the actual objects in the field. Relative accuracy is most frequently expressed in ± inches or feet. For example, a measurement on a map from a water valve to the street centerline must be within a certain relative accuracy requirement to be accepted. Note that relative accuracy does not relate to a reference grid and that the correct geographic position of the object is not relevant. Typical GIS mapping specifications call for relative accuracies of from ±6 inches to ±10 feet.

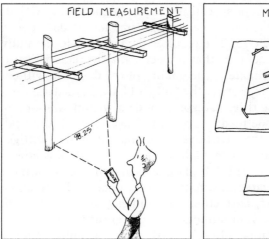
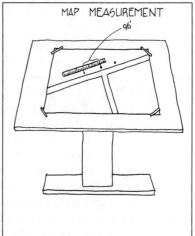

The concept of relative accuracy.

Absolute Accuracy

Absolute accuracy is a measure of the maximum deviation between the location where a feature is shown on the map and its true location on the surface of the earth. Absolute accuracy is generally expressed as a range of inches or feet. For instance, USGS 7.5 minute quadrangle sheets, which are at a scale of 1"=2,000' (1:24,000), have an absolute accuracy of ±40 feet (see section 7.4.1), meaning that most features on the map can be expected to be within ±40 feet of their actual location on the earth. This ±40 feet in ground units corresponds to ±1/50 of an inch in map units. The actual location on earth is measured against the state plane or UTM grid on the quadrangle sheets. This means that most features will be shown at their correct latitude and longitude within 40 feet or less from their true position. Typical GIS mapping specifications call for absolute accuracies from ±1 foot to ±100 feet. Figure 7.1 depicts the concept of absolute accuracy, in which an object is held by its coordinates on the ground and not relative to other objects through distances.

Graphic Quality

Graphic quality refers to the digital map portion of a GIS database and is a measure of the legibility of data, consistency of graphic representation, aesthetics, and adherence to graphic standards usually established as requirements within GIS project specifications. These requirements must be satisfied during the conversion of nodal (point), linear, and polygonal elements that represent features in a vector-based GIS implementation. Nodes, lines, and areas must be converted correctly through adherence to the appropriate symbol, line style, area fill, text placement, and font selection.

Placement requirements are generally documented as standards for the data conversion process to ensure that graphic products created

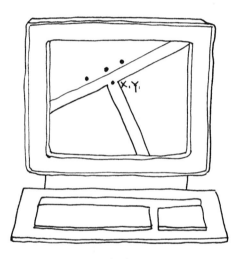

Fig. 7.1. Absolute accuracy.

by the GIS are legible and aesthetically acceptable. For example, a displayable attribute, such as a street name, may be required to be shown on a map inside or outside the right-of-way with the reading angle being precisely parallel to the right-of-way lines. A consistent offset from the right-of-way line may also apply to the street name. Such placement standards allow the GIS users to easily understand and read GIS products and also serve as the basis for the conversion contractor's procedures, software, and training.

Graphic quality also applies to the data conversion of raster images, which is becoming a popular alternative, intermediate, or supplement to traditional or conventional data conversion products. Quality requirements are changing as the technology matures. However, raster images must adhere to specific graphic accuracies that emphasize clarity and aesthetics as the images are repositioned to fit ground control or digitized vector data through warping or rubber sheeting. Cartographic quality also covers the aesthetic representation of objects. It does not look good if two lines that should come together at one point miss each other or seem to overshoot. Apart from the informational quality concerns discussed below, a well-finished GIS map will usually be easier to work with and will find faster acceptance within an organization.

7.3 Informational Quality

Informational quality relates to a GIS database as a whole, including digital map components and attributes. For the purposes of this chapter, topics will be divided into the following issues:

- Completeness
- Correctness
- Timeliness
- Integrity

Completeness

Completeness is a measure of the degree to which all features are included in the database as a result of data conversion. For example, a map that is supposed to show 200 parcels is only 75 percent complete if it shows only 150 parcels. All required features, as well as all attribute data needed to support the GIS, are measured to determine completeness. A typical GIS facilities conversion specification concerning completeness of the deliverables might read as follows:

> All facilities depicted on the conversion sources must be captured. Not more than one percent of the features and attributes, per map, shall be missing from the deliverable data.

The above specification is just an example; acceptance criteria are applied to graphics, attributes, annotation, connectivity, topology, and so on. The price increases dramatically any time an acceptance criterion calls for anything above 90 percent.

In most cases, it is not possible to find a 100 percent complete data source because the real world changes daily, if not hourly. Also, it is rare for historic source data to have been compiled 100 percent complete. Therefore, completeness must relate to source maps and records, not to the real world.

Correctness

Besides being complete, a GIS database must also be correct. A map that substitutes a parking meter symbol for a fire hydrant may be accurate in that an object may be shown at its true location, but the data concerning the type of object is obviously incorrect. Typically, both the conversion contractor and the client will execute automated validations to ensure that the database correctness is acceptable. Attribute values are validated to verify compliance with a specified level of correctness.

Validation can be done manually or automatically; usually, a combination thereof is used. The mix of validation methods used on a given GIS data conversion project is dependent on the data sources. For example, a GIS data conversion specification may require that 97 percent or more of the street addresses be correct as validated against a real estate file. To verify this accuracy, the real estate file will be compared to the GIS file to determine percentage of matching addresses. If there is a 97 percent or greater incidence of perfectly matching addresses, the file would meet the stated accuracy requirements.

Each data feature, attribute field, and database relationship represents a potential for error during the GIS data conversion process. A typical GIS facilities data conversion specification concerning correctness of the converted data might read as follows:

> All facilities must comply with the graphic and database relationships specified herein. Not more than one percent of the facilities and attributes, per map, shall be incorrectly represented.

Timeliness

Timeliness is also a measure of correctness. Data must be current or of a specified vintage. Within the context of conversion, the timeliness of GIS data starts out (immediately upon delivery to the client) as a reflection of the currentness of the source documents used during conversion. It becomes the client's responsibility to maintain the currentness of the GIS database from that point on.

Integrity

A measurement of database integrity is essential to demonstrating and confirming the usability of the data. In a graphic sense, integrity relates to the precise connection of all the lines whose end points have to be collocated, without overshoots or undershoots. In a topologic sense, the database has to indicate which lines are interconnected and which attributes belong with which object (both are aspects of connectivity).

In a database structure, there should be no missing or duplicated data fields. All database pointers (100 percent) must be valid; all records must be correctly built; and the physical database design must be correctly implemented. The GIS data conversion specifications usually indicate that no errors are acceptable in this category.

Clearly, any measure of data quality must include not only measures of cartographic quality but also measures of informational quality. Before informational quality can be measured, a baseline must be established within the GIS data conversion specifications to determine the standard that the database must meet. Collectively, these aspects of informational quality comprise the acceptance criteria for the converted data. Such criteria are normally described in detail within the GIS data conversion specifications.

7.4 Measuring GIS Positional Accuracy

Positional accuracy defines how closely the coordinate position of objects in the GIS database compares to the actual coordinate position of those objects on the earth. Usually, the cost to build the GIS land base increases exponentially relative to the positional accuracy requirements. It is imperative that the positional accuracy be sufficient to support all of an organization's GIS users; however, unique positional accuracy requirements must be carefully analyzed to ensure that the additional cost will be offset by corresponding benefits. Table 7.1 shows the typical map scales used by various GIS client types (market sectors).

Accuracy and Scale

Cartographic quality can be measured according to National Map Accuracy Standards, established in 1941 and revised during the 1940s, that relate to the absolute accuracy of published maps. These standards specify the maximum error that can be expected as a result of the compilation and publication process, and vary the amount of tolerable error based upon map scale. The allowable error varies because the value of this error is directly related to the scale at which the map was originally compiled. Table 7.2 indicates typical accuracy requirements by market sector.

National Map Accuracy Standards specify that for maps of publication scales larger than 1:20,000, not more than 10 percent of the points tested shall be in error by more than 1/30 inch, measured on the output scale, whereby output scale is equal to the scale of the source maps used for input. For maps of publication scales of 1:20,000 or smaller (e.g., 1:100,000), not more than 10 percent of the points tested shall be in error by more than 1/50 inch. Map accuracy depends directly on the publication scale, as shown in table 7.3. Thus, the compilation of maps at two

Table 7.1. Map scale usage matrix.

Market Sector	Typical Map Scales Used (In Feet Per Inch)					
	<50	100	200	400	1,000	>2,000
1. Public Sector						
- Local Government						
- Tax Assessment		■		■		
- Planning			■	■	■	
- Public Works/Transportation	■	■	■		■	
- EMS/Police/Fire	■	■	■		■	■
- State Government						
- Natural Resources						
- Environmental Management						■
- Highways/Transportation	■	■	■	■		
- Lands/State Planning		■				
- Federal Government						
- Agencies						
- Defense	■					
2. Regulated Sector						
- Electric Utilities	■	■	■			
- Gas Utilities	■	■	■			
- Telephone Utilities						
3. Private Sector						
- Forestry			■	■		
- Mining						
- Financial/Insurance				■		
- Real Estate/Title		■	■			
- Oil				■	■	
- Transportation				■	■	■
- Retail/Market Research			■			
- Cable TV			■	■		

Table 7.2. Accuracy requirements matrix.

Market Sector	Typical Accuracy (In Feet)					
	<1	2	5	10	50	>100
1. Public Sector						
- Local Government						
- Tax Assessment		■	■			
- Planning			■	■		
- Public Works/Transportation	■	■				
- EMS/Police/Fire				■	■	
- State Government						
- Natural Resources					■	■
- Environmental Management					■	■
- Highways/Transportation				■		■
- Lands/State Planning					■	■
- Federal Government						
- Agencies					■	■
- Defense					■	■
2. Regulated Sector						
- Electric Utilities				■	■	
- Gas Utilities		■	■	■	■	
- Telephone Utilities					■	
3. Private Sector						
- Forestry		■			■	■
- Mining		■			■	■
- Financial/Insurance					■	■
- Real Estate/Title					■	■
- Oil					■	
- Transportation					■	
- Retail/Market Research						■
- Cable TV					■	

Table 7.3. Relationship of compilation scale and absolute positional accuracy (based on National Map Accuracy Standards).

Map Scale	Accuracy
1" = 50'	± 1.67'
1" = 100'	± 3.33'
1" = 200'	± 6.67'
1" = 400'	± 13.33'
1" = 2,000'	± 40.00'

different scales (e.g., 1"=100' or 1:1,200 and 1"=400' or 1:4,800) from the same types of source materials and using the same procedures will result in stated positional accuracies of ±3.33 feet and ±13.33 feet respectively.

Positional Accuracy Measurements

The GIS data delivered by a conversion contractor are usually checked for data quality so that the delivery can be accepted or rejected. Informational quality can be assessed by visual inspection, and data integrity can be verified with validating software run against the GIS database. But the testing of positional accuracy, relative or absolute, usually requires an external test.

It is usually not possible to test the positional accuracy of every object in the database because this would be more expensive than the data conversion itself. Therefore, statistical methods have to be used to assess database positional accuracies. The testing of positional accuracy has to be completed carefully so that the results have statistical meaning and so that a possible product rejection is credible and defensible.

In the case where a GIS land base is converted from existing map and record sources, the tests have to compare the sources with the resulting database. Proponents of aerial photography point out that using a variety of sources makes checking the GIS database against the sources a more complex undertaking. In the case where a new GIS land base has been created, such as through aerial mapping, the tests can be run against the new sources (the aerial photography) or against the real world through the use of land survey crews employing various technologies.

In any case, the statistical approach is to test a sample of GIS land database points against the corresponding points on the map sources or on the ground. If the conversion of existing maps is tested, a set number of points on the map are digitized, and the coordinates are compared with the coordinates of the same points in the GIS database. If aerial mapping products are tested, ground control land survey coordinates of specific points are compared to the coordinates of the same points in the GIS database. Alternatively, aerial mapping products can be tested by comparing newly digitized photograph points to the corresponding GIS database points.

Statistical testing for absolute positional accuracy consists of choosing a sample of points to be tested. Typically, this sample can be 10 points per map, 20 points per aerial photograph, or 5 percent of all the points in the database. The number of points can be stated in the GIS data conversion specification. The test itself consists of directly subtracting one set of coordinates from another, thereby obtaining a set of coordinate differences which can be combined into a root mean square (RMS) value for the map or photograph. This value should satisfy the maximum error limits stated in the GIS data conversion specification.

Statistical testing of relative positional accuracy can use a similar approach; however, relative positional accuracy calculates differences of measured distances instead of differences of coordinate position.

Reproduction, Scale, and Accuracy

Maps are often reproduced at scales that are different from the compilation scale. The process of enlarging or reducing maps during GIS database output does not change the inherent accuracy of the GIS database, but errors are enlarged or reduced with the map output, relative to the scale of the map that was used as the source. National Map Accuracy Standards suggest that an enlargement of a map, drawing, or published map should state the conditions of the enlargement. The cartographic accuracy standards adopted by the U.S. government 50 years ago were based on manual map production and conventional publication techniques. The advent and general acceptance of GIS technology and precision publishing techniques is leading to revisions to those standards to reflect the differences in compilation techniques and the concept of producing maps at user-defined scales from a single continuous and scaleless GIS database.

7.5 Typical GIS Positional Accuracy Requirements

Determining the appropriate level of positional accuracy requires a good understanding of the intended uses for the GIS data. Often, the need for cartographic accuracy is less of a priority than might be anticipated. While data at a higher degree of positional accuracy may be desired, existing information for features may limit the attainable converted positional accuracy. If a higher degree of positional accuracy is required than current sources allow, a field survey operation incorporating GPS, aerial mapping, or other means of establishing precise coordinate location may be appropriate.

In some cases, issues such as data completeness and currentness are more important to users than positional accuracy. For example, some telephone companies are much more interested in the integrity of the lines in a wire center than in the positional accuracy of the underlying land base. In other cases, positional accuracy is of great concern, such as a gas company's interest in the location of buried facilities under a street intersection.

Assuming similar levels of completeness and currentness, GIS positional accuracy needs can be generalized based on intended uses as shown in table 7.4. Obviously, projects that combine several uses must adhere to the higher positional accuracy requirements.

Table 7.4. Typical GIS positional accuracy requirements.

General Activity	Conceptual Accuracy Requirement
• Tax Mapping	± 2' - 5'
• Final Engineering Design	± 1' - 10'
• Conceptual Plans	± 2' - 100'
• Preliminary Facility Layout	± 2' - 10'
• Vehicle Routing	± 10' - 50'
• Master Plans	± 20' - 100'
• Urban Planning	± 5' - 10'

7.6 Why Positional Accuracy Is an Issue

Positional accuracy is always an issue during the planning of a GIS database because it is opposed by cost. The conversion of data with high positional accuracy is an expensive and time-consuming process. For example, several assessors' offices have attempted to create a precise record, some with a positional accuracy of one hundredth of a foot, for all the parcels in their counties. These projects have been largely abandoned because of the time and resources they were absorbing.

During the study stage for virtually every GIS implementation, positional accuracy is a major issue. Naturally, the positional accuracy of a GIS database and associated costs should be synchronized with the needs of the organization. Achieving 100 percent graphic and attribute data quality at ±1 foot absolute positional accuracy will be very expensive. However, if informational quality and positional accuracy requirements yield a greater payback, significantly reduce the payback period, or considerably increase GIS participation, they are justified. Various combinations of positional accuracy, completeness, correctness, and database integrity should be specified for a GIS database data conversion effort in order to balance cost with requirements.

Table 7.5 illustrates the difference in the cost of producing a GIS land base at varying levels of positional accuracy; it is meant to illustrate relative costs in a very simplified fashion and is not intended to be used to determine actual GIS land base data conversion costs.

Table 7.5. Positional accuracy requirements vs. GIS data conversion costs.

Accuracy	Cost
± 10'	$X
± 5'	$4X
± 1'	$16X
± 1'	$160X

Note that some cartographers would question whether the ±1 inch positional accuracy is even achievable in the context of typical GIS map scales and the original data capture techniques employed to achieve these scales, but precise calculations and large numbers of ground control points can result in this level of positional accuracy.

When determining the optimum balance between GIS data conversion expenditures and GIS data positional accuracies, an organization must also consider the potential for data exchange. If the converted GIS data will be compared to, exchanged with, or referenced to data produced by or belonging to other organizations, this will need to be considered during the development of the GIS data conversion specifications.

There is an emerging trend toward multiparticipant GIS projects. In these instances, several organizations (e.g., municipal government, county government, electric utility, gas utility, and local telephone company) will decide to share in the costs of developing a common land base. While it is often a very demanding process to get these diverse interests to establish a set of common specifications, the result is usually a set of GIS data conversion specifications reflecting the justifiably highest positional accuracy requirements. In this way, all participants are assured that their positional accuracy requirements will be met within reason, and each participant's funding share will be proportional to their database content and positional accuracy demands.

Beyond the conversion cost issue, the ability of the client to maintain the data to the same degree of accuracy becomes an issue. By not doing so, the client will degrade the database over time.

7.7 Verifying GIS Data Quality

Geographic information systems data is verified through a number of methodologies appropriate to the data sources, database specifications, and data conversion approach utilized. Typically, a combination of manual and automated procedures is used to verify positional accuracy, completeness, correctness, and integrity.

Specific to GIS is the problem of verifying graphic and nongraphic (tabular) data in a spatial environment. As data may originate from several sources to be compiled into a GIS, a data conversion checkpoint approach to quality control is common. In this approach, various checkpoints in the data conversion process are identified where GIS data quality can be validated against the source material, database schema, and other standards.

Automated QA

Automated QA requires the use of custom software designed to test the integrity of the converted data. The following checklist includes individual validation checks that are required for most GIS data conversion projects:

1. Each GIS database record must represent a land base or facility feature as specified in the GIS data conversion specifications document (for example, proper layer, symbology, color, etc.).

2. Each land base and facility feature record must contain values for all required attributes, including those specifying feature relationships (where applicable).

3. Land base and facility feature attributes must contain acceptable or default range and domain values.

Data delivery/QC.

4. Attributes containing calculated values must correspond to the values resulting from the application of the appropriate formula or algorithm.

5. For feature records requiring network connectivity, all attributes describing relationships with other network components must be logically consistent. For instance, a 4-inch pipe cannot be directly connected to a 3-inch pipe without the presence of a reducer.

6. For applications requiring connectivity between specified graphic elements, each element requiring connectivity must be connected to another acceptable graphical element. For instance, in a data layer specifying soil coverage, each boundary line between soil types should close to form a polygon.

Manual QA

Manual QA involves processes primarily designed to verify the accuracy and completeness of the GIS output products. Each of these processes generally requires access to check plots of a converted area of interest. In some instances, field verification and/or measurements are required.

To conduct the manual process, the quality assurance technician will compare GIS database check plots to the source documents to verify that the following requirements have been met:

1. Absolute/relative positioning must be within acceptable tolerances.

2. Land base and facility feature annotation must correspond to the information provided on the source document and be in accordance with the graphic specifications.

3. Text offsets and orientations must be correct and be in accordance with the graphic specifications.

4. Land base and facility feature representations must correspond to the source document representations and all must be present.

5. Features spanning multiple plots must be correctly edge matched.

6. Obsolete source record symbology must be correctly translated into GIS standard symbology in accordance with GIS data conversion specifications.

Field verification or measurement may be appropriate to verify that the following requirements have been met:

1. Converted land base and facility features must be positioned on the GIS check plot within acceptable tolerances for relative and absolute accuracy.

2. Land base and facility features must be classified correctly as determined through field observation.

3. Land base and facility feature displayable attribute values included on the GIS check plot must correspond to those values observed or measured in the field.

4. Land base and facility features locatable in the field and specified for data conversion must be included on the GIS check plot.

5. Land base and facility feature attributes must be converted whenever the attributes are observable or measurable.

(NOTE: Attributes requiring verification do not always appear on graphic check plots as displayable annotation. In these instances, it may be necessary to print some portions of the GIS database in a report format suitable for assessing the quality of the data conversion process [e.g., a listing], or to interactively check attribute values at a workstation or terminal.)

Additional check plot processes are conducted to verify that the following requirements have been met:

1. Land base and facility features depicted on the source documents and specified for conversion must be included on the GIS check plot.

2. For each feature depicted on the source documents for which annotation is required, feature annotation must be converted, formatted, and displayed in accordance with the GIS data conversion specifications.

3. Default attribute values must only be used in feature annotation when an acceptable attribute value does not exist on the source documents.

4. Graphical connectivity must be maintained in accordance with specifications.

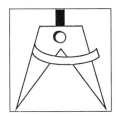

Chapter 8

GIS Database Design

8.1 Introduction

Most GIS have database management functions inherited from the system software they are based on. As such, GIS cannot generally be used without the prior design of a database structure (i.e., database schema). Therefore, database design is a necessary prerequisite to conversion primarily in terms of satisfying applications requirements. The two design phases of a database are identifying the user requirements, and applying the requirements in a physical database design. These two steps are based on a significant amount of analysis, design, and construction activities, each requiring considerable expertise.

The GIS database design, in its initial implementation, usually relates to one specific project, for which only one database structure will be needed. But a GIS can generally handle more than one project and database. Therefore, more database structures can, in the long run, be designed to suit a user's different data requirements for different GIS projects. When each database is called up, the corresponding database structure is called up automatically by the GIS.

To effectively design a database, an enterprisewide view of the data requirements is needed to avoid data duplication. The designers must understand the overall business processes that the data will be supporting.

8.2 Database Design Stages

Database structure is common to all three design stages. First, the structure is created conceptually, then as a clear document specifically directed to a physical and vendor-specific GIS, and finally in the form of an actual implementation in the GIS memory. Figure 8.1 illustrates the database design stages. These stages are described further, but in the remainder of this chapter the differences among them are no longer stressed.

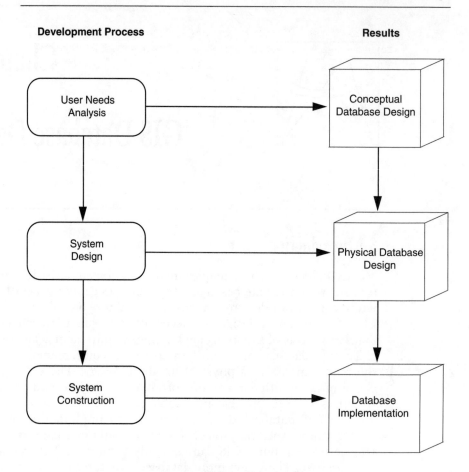

Fig. 8.1. Database design stages.

Conceptual Database Design

During this first stage, the database designer typically focuses on the content of the GIS database rather than on its structure. In cases where special structure requirements have been identified, these are also collected and documented.

An important part of the conceptual database design is listing the GIS database elements. These define the objects that the database has to provide. For example, if the database has to contain parcels, then the database design has to specify the presence of this specific type of data. Since many parcels are expected, the database design has to establish a feature class or object called parcel (note the singular form). A specific parcel then is one feature or event of that feature class.

The first step in the conceptual design stage is to clearly identify the organization's GIS data needs. Needs assessments may be compiled from interviews and surveys of potential GIS users who provide input regarding the data and functionality required to perform the full range of anticipated GIS applications. A design is then developed that logically presents the GIS functional and data requirements.

Requirements associated with the implementation of a vendor-specific GIS are often intentionally excluded from this stage of the design. Doing so ensures that the conceptual database design is not biased toward a specific vendor, especially since portions of the conceptual design are often included in requests for proposals and therefore precede GIS vendor selection.

Physical Database Design

During the physical database design stage, the actual structure of the GIS database is developed and documented based upon the content (features and attributes) identified in the conceptual design stage, but now targeting a vendor-specific GIS. Therefore, the physical design is done after a GIS vendor has been selected.

While the physical database design stage may also involve changes to the content of the GIS database, these changes are typically driven by the need to accommodate the vendor-specific GIS and DBMS that were selected. Physical database design cannot occur without knowledge of the vendor-specific GIS and DBMS which will be used to manage and access the data.

This process has been, to date, a predominantly manual task. Experienced GIS database designers are in great demand. Computer-aided software engineering (CASE) tools are just now making inroads into the GIS arena.

Database Implementation

The last stage of the GIS database design process involves the actual coding of the physical database design. The physical design stage generally results in a paper document suitable for distribution, review, and approval. Once GIS user input is received as a result of the review process, the final paper version of the physical database design is ready to be entered on the GIS in the form of a project-specific GIS database schema.

8.3 What Drives Database Design?

A multitude of factors influence the content and structure of the database used to support a GIS project. While many of these factors can be easily investigated and accounted for in the design of a GIS database, others may require a more thorough analysis of the environment in which the system will be used. The following sections discuss the major forces that drive database design.

User Needs

Geographic information system databases are usually developed to serve the needs of users within the scope of a certain project (which can be rather broad). Lists of potential GIS requirements (including applications requirements) are compiled through interviews and surveys or through user representatives who contribute subject matter expertise. Multiparticipant GIS projects are characterized by diverse requirements. These requirements are analyzed and categorized into data types that support the functional requirements of the GIS. The degree of database intelligence (see page 11, "Geographic Information Systems [GIS]") is determined through analysis of the data relationships necessary to support the GIS functional requirements.

During the analysis process, shared GIS user requirements are identified to determine where the data will originate and who will maintain the data. Centralized versus distributed approaches to GIS data access are analyzed. Functional requirements are translated into application acquisition or design. User environments specify GIS requirements as to potential interfaces and compliance with operating standards. Output product needs (as identified during the functional requirements definition stage) may impose additional requirements to support hard copy reports and other noninteractive GIS activities. All these requirements have to be collected into a comprehensive report, as shown in figure 8.2.

Ultimately, analysis and design produce a logical data model that will support the GIS user requirements. It is recommended that a structured methodology be used to develop a GIS database design. This approach involves a team effort to complete a baseline GIS database design on paper, followed by the detailed design being generated and coded into a digital schema, and culminating in a team review of the end product on the GIS.

Fig. 8.2. User needs survey.

GIS Application Data Requirements

In many cases, GIS users are able to describe the tasks they need to perform without being able to specify the GIS data items required to accomplish the tasks. For example, GIS users may describe an application that generates mailing labels for every property owner located within a given radius of a specified parcel. The most straightforward implementation of such an application will involve polygon processing and will, therefore, require that parcels be created as polygons rather than as lines. This application will also require that the owner name be stored as a parcel-related attribute rather than as an unintelligent annotation on a map.

Applications may be obtained directly from GIS vendors as custom code or off-the-shelf software. It is usually prudent to identify existing GIS applications that may meet user requirements as opposed to paying for new development. Off-the-shelf applications will, to some extent, determine GIS data requirements. For example, a utility transformer load management application will require that specific electrical network components reside in the GIS database. The design will have to include the data requirements to support such applications.

In the event custom application development is required, GIS database design must take into account the data requirements of the application. Again, use of a structured methodology for the development of GIS applications is recommended.

Available Data and Cost of Conversion

Ultimately, the GIS database will contain only that data which is either readily available or which can be compiled economically to satisfy user requirements. In effect, it is necessary to perform a cost-benefit analysis (albeit a very simple one) on each item (or class of items) to be included in the GIS database. The questions to be answered include the following:

1. What is the cost of acquiring the data item?
2. How many GIS users need the data item?
3. What is the cost of maintaining the data item?
4. Can the users' needs be met with the use of a substitute data item?

Subsequent reduction of the data set because of conversion cost restrictions or lack of source information is common. Careful synchronization is required to maintain forecasted benefits while redesigning the GIS database.

Naturally, the conversion strategy will avoid the cost of converting data that may already exist on another system. Both the conceptual and physical database design processes have to take into account interfaces with other existing or planned computerized data, including the structure and type of that data so that they can be brought into the GIS database, if required. In considering the influence of interface requirements, three distinct possibilities exist regarding the amount of data to be stored in the GIS and the direction of the data flow.

1. In some cases, the other system is considered an external data source, and it retains the responsibility for data maintenance. Data are requested by the GIS and transferred from the other system as needed. The transferred data may be stored on the GIS (even if only temporarily), or they may be directly displayed on the GIS, such as a raster image of a graphic record in a separate display window.

2. The other system may be considered an external data source and destination. That is, the GIS requests data from and transfers data to the other system. In this type of interface, database maintenance responsibility can reside with the GIS users (as opposed to the users of the other system).

3. A one-time bulk load of data from the other systems or data sources (e.g., TIGER files) is entered into the GIS, either as a separate file or incorporated into the GIS database. Again, ongoing data maintenance responsibility resides with the GIS users. Once the data transfer is completed, the other system may be phased out.

In all these scenarios, the GIS database design must accommodate the match key attributes necessary to request information from other systems, to receive data from other systems, and to hold data to be sent to other systems. In some cases, this requirement will be limited to the inclusion of a common key item to relate the GIS data to the other system. At the other end of the spectrum, and especially in support of the second and third scenarios presented above, the GIS database design will have to include a significant portion (possibly all) of the other system's data contents. If GIS applications are designed to request information from external databases, it may be that the information received is only used by the application and that no special attribute field has to be established.

The complexity introduced to the GIS database design process to facilitate these interfaces is well worth the effort in terms of both reduced conversion costs and overall operational effectiveness.

Data Conversion Schedule

The data conversion schedule affects both the conceptual and the physical database design. During the conceptual design stage, the data conversion schedule will determine when GIS data items (and the applications that rely on them) will be available for use. In some cases, the database designer may be able to substitute an item from another source that can provide some of the functionality needed until the preferred GIS data is converted.

During the physical design stage, the database designer must account for the availability of the GIS data when designing the structure of the data tables or data layers. Instead of designing the entire structure, it may be more practical to design or implement only those GIS data items that will be available in the short-term and to plan for additional physical design tasks throughout the data conversion process. Under this approach, the GIS database is incrementally designed and implemented throughout the life cycle of the data conversion project. The need for an incremental approach has to be identified early so that the selected GIS allows the database structure to change easily, without affecting existing data.

Future Expansion

Geographic information system databases are dynamic. They change and evolve over time in response to new or changing needs. Future expansion must be taken into account during the conceptual and physical database design stages. Accommodating future expansion is not easily done. Obviously, a dilemma exists when a physical database must be implemented before the conceptual design has been completed for all applications or before all interfaces have been identified. Often, the experience of the designer is a major factor in ensuring that the GIS database incorporates the flexibility and versatility required to accommodate future changes.

Maintainability

As discussed in chapter 4, "GIS Data Sources," the topic of database maintenance is separate and distinct from data conversion per se. That is, to fully address the issues related to the ongoing maintenance of a GIS database would require another complete publication. However, in the context of database design, the ease of database maintenance is a very important consideration from the client's point of view. To enhance the chances of project success, it is vital that the database design reflect the end user's perspective in terms of common access methods, routine queries/outputs, and the ability to add, modify, or delete selected items within the database.

8.4 GIS Database Design Elements

A GIS database consists of logic data, graphic data, attribute data, and data relationships (intelligence). The ability to manipulate these elements in a spatial environment provides the basic functionality of a GIS. Relationships between data types provide the more complex information structures that many GIS users require. The ability to trace a network (node), relate a building footprint to a street address, or edge match facets depends on the established relationships of elements in the database. The following section describes the utilization of these data types in a GIS database.

Logic Elements

Logic elements provide the positional (X,Y) reference structure that holds graphic information, and are designated as nodes, links, chains, and areas. Areas are also known as *polygons*.

A *node*, which is nothing more than a point location described by coordinates, can be the center of a hydrant or it can be the locus where two or more lines join. A *link* represents the shortest segment that joins two nodes. A *chain* is a collection of links, where one link connects with the next. An *area* is a two-dimensional space that is surrounded by a closed chain.

As defined, nodes, links, chains, and areas are important constructs in a GIS database. They do not have a graphic representation, unless a graphic element is assigned to them. They are used to define the type of database objects during design, and they are used to help the database management system keep track of the object types.

All logic constructs are built on the basis of nodes, and as such they impart the geographic location and shape to the digital map portion of the GIS database. This makes it possible to move, for example, a hydrant (node). Such an operation updates the coordinates of the node, but the pointers to the hydrant symbol and the attributes are retained. In this example, the hydrant, including the node, the symbol, and related attributes, is called a *feature*.

During the physical database design, every GIS database feature has to be assigned to one of the logic types. For example, a hydrant would be of type node, and a parcel could be of type area. Figure 8.3 depicts the logical correlation between map features and database elements.

GIS Data Conversion Handbook

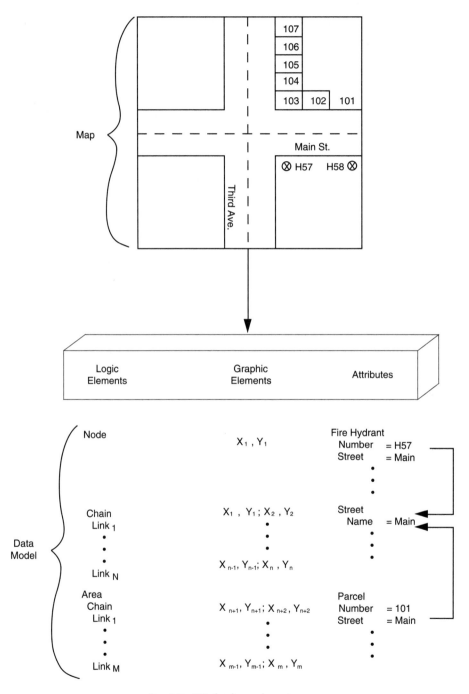

Fig. 8.3. GIS database elements.

Graphic Elements

Graphic elements are used to display features pictorially in a GIS and are held in place by the logic elements so that they are spatially placed in a coordinate location. Graphic elements are symbols, line styles, and area fills, and they carry all the graphic representation information necessary to display the information related to logic elements. This makes it possible to change a symbol or a line style without losing the logic elements.

The design of graphic elements is part of GIS database design. This involves the assessment of existing symbols, line styles, and area fills, and the design of new ones. During GIS database design, assignment of graphic elements is made to those features required to be accessed and maintained in a graphic fashion (e.g., a fire hydrant may be represented by a unique symbol). Data not required to be pictorially accessed are usually stored as tabular attributes that may be linked to a feature.

Typically, physical features are those most often depicted with graphic elements. Features such as parking meters and fire hydrants may be stored as nodes. Features such as street centerlines and water mains may be stored as links or chains. Features such as parcels, for which areas may be calculated, are stored as areas or closed chains (polygons).

Most GIS vendors allow graphic elements to be reassigned freely to different features. For example, the line style for the feature street centerline can be easily changed without affecting the centerline's position or attributes. Today's GIS are commonly designed with a separate database for graphic elements which is linked to the logic elements through database management software and data (lookup table).

Attributes

Attributes are alphanumeric data that are associated with features and are typically stored in separate databases (some systems do integrate graphics and text). Attributes hold a variety of additional information for features. In the earlier example of a fire hydrant, its type, manufacturer, year of installation, size, and date last flushed might each be an attribute.

Attributes may be displayed along with the graphic feature (e.g., a street name) or remain unseen on graphic displays (e.g., a phone number), yet still be accessible via user queries or applications. The physical GIS database design has to include the display rules for attributes.

Attributes may also be used to hold access keys to/from other databases. These keys can be used to find data that are outside the GIS database, in the form of digital data. During GIS database design, each attribute is defined in terms of data field type, format, and length, in the same fashion as fields of a tabular database.

A variety of data types are usually provided by database management systems, such as integer or real or floating point, and character or string. Others offer additional types such as Boolean, date, bulk text, or dollars.

Each attribute has to be assigned to a specific feature class during database design. Therefore, the database management system learns that, for example, all hydrants have an attribute called date of installation which is of type date.

GIS Data Relationships

Three types of GIS data relationships must be determined during the design of the database: relationships between feature classes and their attribute types, relationships among attribute types, and relationships among feature classes.

Relationships between features (e.g., parcel outline) and their attributes (e.g., parcel owner) can be defined using a unique physical pointer of data, which the data conversion operator assigns during the data conversion process (e.g., parcel identification number), or with a system pointer, which the GIS supplies automatically (e.g., unique identification). With the physical pointer, care has to be used to ensure that the identifier is assigned uniquely and systematically. For example, while street addresses can constitute an adequate identifier for parcels, they are often not coded consistently. The GIS database designer has to respect the type of feature-attribute linking tools the database management system is able to offer.

Relationships among attributes are built via user assigned or system defined unique identifiers. These relationships may be defined by tables or, in the instance of an external database interface, a common or match key.

In some instances, however, it may be necessary to use multiple keys to allow multiple databases to be linked. This is particularly true of external databases (databases that are separate from a project specific GIS database, residing either on the GIS or on other systems). The integration or interface of a GIS database with another database is a unique design issue that requires significant analysis. As previously discussed, these considerations are best addressed during the conceptual design stage.

Relationships among feature classes can be established within some GIS. For example, the user wants to create a feature class called voting district which is made up of a few parcel features. This hierarchical relationship between features, where parcel is the child of voting district (the parent), can be established within some GIS packages.

A similar mechanism, called a one-to-many relationship, usually defines a connected network of features that are linked to support a GIS user requirement. If the tools are not provided by the GIS vendor, the hierarchical tree of one-to-many relationships (e.g., water main to water taps) requires that the data conversion process establish many unique pointers in a GIS database.

When all three data relationships are required, a complex GIS database design is the result. Similarly, a complex data conversion approach is usually required to populate the database.

Existing Digital Data

The availability of digital data can have an impact on both the design and the initial population of the GIS database. The need to access or integrate existing data requires that new keys and other unique identifiers complement those used by the existing data. Furthermore, the expense associated with translating or modifying the existing data may require that the data be used in its current (unaltered) form.

Since the issues involved are both abundant and complex, a strategic GIS data interface and acquisition plan should be developed in instances where many databases are to be interfaced to or integrated with the GIS database. Any assumptions about where data resides, the match rate with other databases, the data format, and the benefit of integration and interface should be rigorously challenged before they are included in the GIS database design.

Raster Image Data

The need to include raster image data in the GIS database affects the design of the database in several ways. For example, the large size of raster image data files and the low speed of raster data exchange over a communications network may require that raster image data be duplicated at several locations. This file duplication necessitates the inclusion of appropriate mechanisms to ensure that the integrity of the database is not diminished and that the general management of updates occurs in an effective manner.

The differentiation between GIS that support raster image data and document imaging systems that manage spatially oriented data is slowly narrowing. It is clear that GIS is more and more interlinked with information systems, and in larger organizations it is becoming part of an enterprisewide approach to data management. This development can make database design and resultant data conversion processes even more complex.

Other Data Formats

The evolution of GIS and technology ensures that GIS databases will need to handle a variety of data formats in the future. These formats are first being implemented as attributes (e.g., raster images of valves or transformers) and as whole database solutions (e.g., a seamless raster land base). The foreseeable future holds technological innovations for GIS such as video images and film, satellite imagery, voice and sound, and other static and dynamic data that incorporate vector graphics or imaging, also known as multimedia. Video represents the potential for interactive access to a wealth of visual information that is quite different from stored textual attributes.

8.5 Graphic Structure

The design of the graphic representation of the GIS database features is generally done in two parts, whereby symbology is first created and then assigned to feature classes. The symbology is a combination of graphic constructs using line styles, color, and text.

Symbology

One component of GIS database design is the specification of the symbology used to represent various graphic elements. For instance, a complete description of the graphic representation of a GIS symbol may include the shape of the symbol, its data layer, orientation, color, pattern, pattern orientation, line types, and line width.

The GIS symbology used should take into account the needs of the organization, current symbology conventions, capabilities of the GIS hardware and software, and general appearance on screen or on hard copy. Table 8.1 shows some common GIS symbology for sample node, link, and area features.

Table 8.1. Sample GIS symbology.

Graphic Element	Example	Sample GIS Symbology
Node	Fire Hydrant	⌐○⌐
Node	Water Valve	▷◁
Link	Water Main	▬▬▬▬▬▬
Area	Service Area	— — — — —

Color

Color plays a significant role in differentiating features in a GIS. With line styles being so prevalent in a GIS, color is assigned to line styles to differentiate features visually. It is very common to assign one color to street centerlines, another to curblines, and another to facilities located in the street right-of-way.

While use of color can be implemented quite creatively on a color display screen, it is usually implemented differently for hard copy output since the color handling capabilities of plotters can be limited and because most organizations still copy and distribute information in black and white. This may necessitate separate color assignments for screen display and hard copy output, and if a principal application is based on hard copy output, it can have a bearing on GIS database design.

Color may also play a part in differentiating features with a common theme. For example, it is common to store facility features with different colors based on their construction status (i.e., planned, in construction, and as-built).

Color is often used by conversion contractors to differentiate and clarify data during the data conversion process. The process generally consists of parallel operations that produce independent sets of converted data. These independent sets are then merged into the final GIS database structure prior to delivery to the client. Color helps conversion personnel quickly identify independent data sets and/or the status of conversion prior to the final merge operation.

Color is also very helpful in QA testing. Database audit software can temporarily change the color of all features and/or displayable attributes that fail the test. Quality assurance technicians can then quickly locate and correct errors based upon colors.

Geometric Integrity

Inherent in the overall GIS database design process is the need to analyze the geometric requirements of the database features. Determining which features are to be held by nodes, which by links, and which by areas is not always as straightforward as it may appear at first glance.

Again, the analysis of anticipated applications becomes the driving force behind ensuring that the GIS database design properly accommodates geometric integrity. Take hydrography for example. There is little doubt that single line creeks would be digitally modeled in the GIS as chains. However, rivers (shown with both banks as chains on the source document) could be input as two independent chains (one for each bank) or as an area. The choice will generally be made based upon the needs of one or more applications. In the river example, if area calculations are anticipated as a possible application, the database designer would opt to model rivers as polygons. In a typical GIS database design process, a large number of such fundamental design decisions affecting geometric integrity will need to be made.

Text Annotation

Text annotation visually related to features on source documents has been historically handled within GIS databases in one of two ways: as database attributes that are displayed according to display rules, or as dumb graphics (not explicitly related in the database to the element). Text annotation may simply be placed graphically in the coordinate space and in accordance with the specified graphic parameters (orientation, offset, justification, size, font, etc.), but this has to be designed also. The display of database attributes as graphics annotation often requires an application that supports such functionality. The degree of custom application development varies among GIS vendors due to differences in their systems' architecture and design philosophies.

Typical examples of this issue are street names and street centerlines. Street names may simply be entered graphically in a GIS as text annotation. There may be no direct relationship between the street name and the street centerline with the possible exception of a shared data layer. However, such a GIS database design would not support a query to find the intersection of two streets and may not even support a query to find the text annotation.

Alternatively, one of the street's attributes (the street name) may be earmarked for graphic display under certain display rules. Both earmarking attributes for display and establishing display rules are database design activities.

Layers

Traditionally, GIS graphic elements have been organized in layers. As a simple way of partitioning data, each layer consists of related information, such as all the roads, highways, and trails. Based on these types of layer assignments, selective display of layers would produce thematic views of features data and associated attributes (if desired).

With the advent of object-oriented programming and object-oriented data structures, many GIS software vendors are abandoning the data layering model in favor of an object-oriented data model. In general, it is possible to simulate a layered data organization with objects by requesting a display of selected feature classes.

Visibility Rules

The question of visibility must be addressed for all graphic parameters as part of the project definition process. For workstation activities that involve screen displays, each symbol, line type, text component, and layer needs to be designated as either displayable or nondisplayable. The mix of displayable and nondisplayable items will, most likely, change for specific functions. For example, the display of a primary feeder route for an engineer in an electric utility will generally contain fewer displayable items than a display of the complete distribution system (i.e., primary, secondary, and individual service drops) the engineer might use while designing a work order for an area of interest involving less than a city block.

The same definition process is required for graphic output products such as maps or drawings. Each type of output requires complete specifications for which items are plottable and which are not.

8.6 Topologic Structure

Early GIS databases stored maps in sequential fashion. If the user wanted to find related objects in it, such as two lines that shared a common point, the whole database had to be searched for the second line having the same coordinates for one of its end points. This sequential search was very time-consuming.

To expedite searches, relationships have been built into newer GIS databases that point at related objects. The database knows which line connects to which on the basis of additional data (called *pointers*). Pointers can be seen as extra overhead on the GIS database, introducing a certain amount of bulk. But the improved response time is considerable, and therefore, these relationships (such as locations, residency, coincidence, adjacency, and connectivity) are desirable.

These relationships are collectively called *topology*, and are implemented as pointers in the database and as database management software able to use these pointers. Topology relies on the geometric relationships among logic elements (nodes, links, and areas).

Topology facilitates operations such as finding points in polygons (e.g., which fire incidents were in District 5); finding lines in polygons (e.g., which property lines would be affected by a proposed road widening); and overlaying polygons (e.g., which properties are in a 100-year floodplain).

Topology may also be used for network or connectivity analyses such as determining all water clients affected by a break in a particular section of a water main or determining the shortest route for an emergency vehicle.

Node Topology

In addition to the *X,Y* coordinate value for each node, which defines its geographic location, topology generally requires that a unique identifier be associated with each node. Nodes are also assigned supporting data types which allow them to act as controlling structures in the required topological relationships. For example, the node representing a street intersection can prevent a left turn from a certain incoming link, or a node representing a pump can prevent reverse flow from occurring through the pump during water flow analyses. Also, topologic overhead usually keeps track of how many links are connected to a node.

Note that the specific implementation of topology on various GIS on the market today differs from vendor to vendor, depending on the programmer's approach. For the purposes of this publication, the discussion of topological terms and the reliance upon topology by modern GIS is presented generically to address conversion-related issues.

Line Topology

As previously discussed, a link is a segment that joins two nodes to form a single logic and geometric entity. A *from* node defines one terminus of the line; the *to* node defines the other. Additional *X,Y* locations may be stored for other points (often referred to as vertices) occurring between nodes. In geographic terms, a line may represent a linear feature (e.g., a curbline) or a boundary between polygons (e.g., a county boundary line).

To establish topology in a GIS, links are logically joined at common nodes; the join is usually established by special software. Additional information is attached to each link to store pointers to the surfaces which are to the left and right side of the link.

Area Topology

Area topology is what truly separates a modern GIS from a CAD system. With area topology, a boundary between two areas (e.g., two tax districts) is stored as a series of pointers to existing lines; hence, no redundant data. In nontopological systems, each polygon would be enclosed by its own boundary line, thus creating two common lines along the boundary. In a topological system, such applications as perimeter calculations and area calculations are still supported, since the lines the polygon points to form a closed chain. Database design must declare that a feature class is of type area for these mechanisms to be activated in the database.

Attribute Topology

Attribute topology basically consists of the link between a feature and any one of its attributes. This is usually established on the basis of a database pointer to the information location in the database. Topology greatly accelerates information retrievals, in the same way that line connectivity is used to quickly find the next connecting line without searching the whole database for similar node coordinates. The link between an attribute type and the corresponding GIS feature class has to be established during database design. Once the link is established, specific attributes can be linked to corresponding features during data conversion.

8.7 Tabular Structure

Most modern databases are built using tabular data. There is a variety of implementations of tabular structures, and some utilize hierarchical data structures to create an object-based approach. Simpler, less complex data structures such as flat files are being used also, but only for less robust GIS functions.

Flat Files

Flat files or list-oriented databases are the simplest, though not the most efficient, way to store data. Flat files organize data in sequential lists. These lists are easy to create and view but do not provide any way to perform speedy searches. Locating data in the file typically involves slow searches through the entire file. Databases for GIS do not use this approach except to populate portions of the database with information, such as textual descriptions of events.

Hierarchical Model

Hierarchical databases were, until a few years ago, widely used in GIS. The hierarchical database model establishes parent-child relationships between various attributes in the database. For example, a property owner in a parcel database might be the parent of one or more parcel records as shown in figure 8.4. Each parcel might, in turn, be a parent for zero, one, or more building permit applications.

This structure allows easy access to data, provided the appropriate information is known. For example, if the owner name is known, the system can easily and quickly locate all properties owned by that person and all permits applied for on those properties. However, finding the owner who applied for a building permit is a more lengthy process.

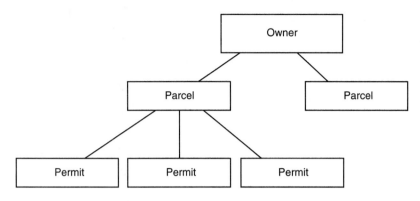

Fig. 8.4. GIS hierarchical data model.

Relational Model

The relational GIS database model has increasingly gained support and popularity in virtually all sectors. Among the reasons for its success are the simplicity of the underlying concepts, the versatility of the resulting databases, and the ease with which data relationships can be established. Because of the ease of creating new relationships, the design and review process can be shortened.

Relational databases organize data in tables (also called records). Each table is made up of rows (also called tuples or relates) and columns (also called fields or attributes). Information is usually organized in multiple tables that share common keys as illustrated in figure 8.5.

The majority of the relational database management systems available on the market today support the Standard Query Language (SQL). The SQL gives a great deal of flexibility in constructing ad hoc queries. For example, selecting the address of a parcel whose owner is named "Smith" is as simple as:

```
SELECT address FROM parcel WHERE owner = Smith
```

Because of its level of standardization, SQL is becoming an essential component of GIS databases in which various pieces of GIS data may reside on different computers running database management software from multiple vendors.

The query languages supported by the GIS vendors will form the basis for application development, and as such their capabilities will have to be assessed during the physical database design and actual database structure implementation. Without this knowledge, applications may not be able to properly satisfy data extraction requirements.

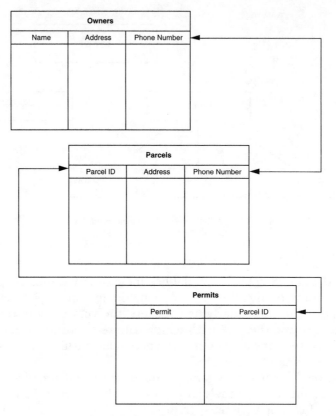

Fig. 8.5. GIS relational data model.

Conversion contractors have embraced relational databases as tools for data conversion due to their flexibility and ease of use. Relational databases allow conversion contractors to easily format attribute data for a variety of vendor-specific GIS.

8.8 Interfaces

Most GIS implementations require some degree of interface development. The most effective GIS installations are those where the system does not operate on a stand-alone basis. As mentioned previously, many organizations will already have a significant investment in digital data on other systems. For the GIS to be most cost-effective, it must be able to transport data to or from these other systems, or, ideally, access the data and functionality of these other systems within the operational framework of the GIS. The structure of the external databases to be interfaced with the GIS will have to be defined for integration; this information needs to be made part of the database design.

Chapter 9
Multiparticipant Databases

9.1 Introduction

This chapter is designed to introduce the reader to an important database assembly solution that is based on sharing portions of the cost to create a common GIS database among two or more legally separate organizations. If a group of two or more organizations decide to share database portions and data acquisition costs, then the process becomes known as a multiparticipant project.

A more refined definition of a multiparticipant project is difficult to make because the relationships among the participants can take many forms. But the principal characteristic is the sharing of data in order to reduce costs for all participants. For example, a city, a county, two utilities, and a telephone company may share the cost of developing a land base even if each may separately acquire other database components that are not shared.

Multiparticipant projects introduce opportunities to change the way business is conducted. New methods of data exchange, and organizational and technical information interfaces, along with more responsive computer capabilities, mean that business methods may change after project implementation. Innovative groups recognize this potential for change and increasingly take steps to restructure their organizations around the use of new GIS databases in order to improve services and cost-effectiveness.

Participants are usually organizations from the public and private sectors. The interest in data sharing has frequently resulted in legal relationships among local governments, public and regulated utilities, universities, and other commercial organizations. Figure 9.1 shows the typical organizational structure used for multiparticipant projects being created to share common GIS data.

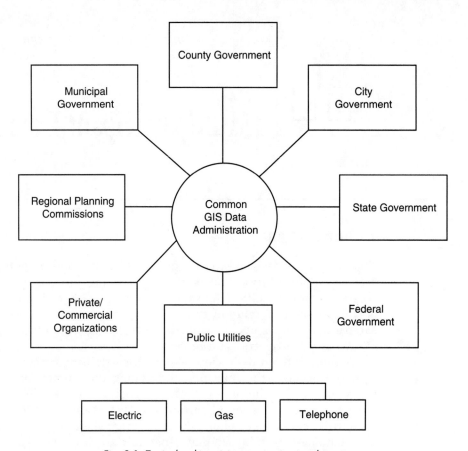

Fig. 9.1. Typical multiparticipant organizational structure.

Several developments have contributed to the growth and acceptance of multiparticipant projects. These include network communications, open data architectures, and the availability of system integration services. Also, ratepayers, taxpayers, constituents, and shareholders are pressuring organizations to demonstrate fiscal responsibility by finding creative solutions to integrate and use common information. The costs for daily operational tasks, such as cross-checking paid tax or permit fees or determining underground utility locations, can be significantly reduced when data is shared.

In a given geographical area, the clients of utility companies are also taxpayers. It makes fiscal sense to them not to have these utilities and the government fund the creation of two or more separate GIS land bases of the same area and of the same type, when a common land base could be shared by all the parties.

The public is beginning to demand that service organizations exchange information. The exchange of information is difficult at present, especially when the organizations have not previously done so. But, as technology evolves and multiparticipant projects mature, data exchange will become much easier.

It is the cost savings associated with land data capture and ongoing maintenance and exchange of this data that promotes the start-up of multiparticipant projects. For example, a participant in a midwestern multiparticipant project was able to acquire a high-quality land database for less than a quarter of the cost that the company would have spent on data conversion if it had proceeded independently. Therefore, even if each project is different, and even if organizations are affected by different regulations, competition, and constantly evolving business practices, the establishment of a multiparticipant project can be a prudent and very economical business decision if the GIS solution allows for properly planned and implemented sharing of data.

A multiparticipant project can only be established if more than one organization is interested in obtaining GIS data within a specific geographic area. The identification of potential participants is an important preparatory step prior to any actual database work. It is easier for a multiparticipant project to get started when initial membership is limited to those organizations that have the commitment, resources, and funding for the always-challenging start-up period. Other interested organizations should be contacted and a framework must be established to allow them to join or participate later. Sometimes, subsequent participants join with a different fee schedule or modified contractual terms to suit their funding. The criteria that determine an organization's participation in a multiparticipant GIS project are as follows:

- Geography
- Cost
- Politics
- Cooperation
- Interest
- Funding
- Specifications

9.2 Standards

The more participants a project has, and the more complex the proposed database turns out to be, the more critical the definition of expectations and standards becomes for a successful technical implementation and cost-effective use of the multiparticipant database.

Specific project standards associated with database design, accuracy levels, graphic presentation, data exchange, and communications must be determined. This process will usually require extensive evaluation efforts.

9.3 Organization

The organization of a multiparticipant GIS project is usually much more complex than a single organization project because of the variety of interests, schedules, budgets, and project requirements that are present. Each participant has to have the appropriate level of input in the project, and supervisory and technical entities have to make sure that all interests are appropriately administered.

The Executive Steering Committee

Initially, an informal group is set up, which over time, becomes better organized and more formalized. To administer its activity, the group may develop a separate legal entity or it may be based on a loosely organized board of directors in which every participant is represented. In general, this executive steering committee will consist of administrators from the various participants (e.g., a city manager and executives from utilities).

The number of executive steering committee members has to be limited to a manageable number and, of course, to those participating entities. Most projects in North America include representatives from a county government (in the United States) or a regional government (in Canada), a principal city, and the major utilities within the defined geographic area. In many instances, smaller municipal governments within the defined geographic area also participate.

At the executive level, issues such as implementation funding, data ownership, legalities, and multiyear arrangements (i.e., maintenance funding) are discussed and settled. The executive team also provides management support, vision, and guidance to the technical team (see page 171, "The Technical Project Team") by sharing most decisions with it.

The Project Champion

Virtually all successful projects have a project champion. This individual, typically an executive in one of the organizations, is characterized as someone who has a great personal interest in the implementation of a GIS and who is either familiar with the technology or is actively learning about it. He or she is often able to secure executive-level interest and support for a GIS project, based on technical practicality and economics, from one or more organizations. The champion identifies and recruits other potential participants.

The Technical Project Team

Once the principal participants are identified, a technical team consisting of technical representatives from all the participants is formed. This group includes direct GIS users such as heads of engineering and planning departments.

At the technical level, issues associated with source information preparation and scrub, database content, organizational structure, system architectures, data transfers, maintenance procedures and schedules, and day-to-day communications are defined. These are documented and presented to the executive committee for approval.

The Project Manager

In addition to the two committees and the champion, a qualified project manager is another key to a successful multiparticipant project. The project manager manages the technical team, is responsible for coordinating the major functions of the project, and acts as a liaison between the technical team and all the participants. The project manager also handles external activity with conversion vendors, system vendors, and consultants, and as such has to have a very good understanding of each participant's map and record processes and data requirements.

9.4 Agreements

Whenever diverse groups such as local governments, utilities, and private commercial entities become involved in a multiparticipant GIS project, some form of written agreement is necessary to protect the interests of all parties to the extent possible. Generally, the feasibility study and preimplementation portion of the GIS project require one kind of agreement and the implementation phase requires another more formal and detailed one.

A master agreement is typically used to define the participant's commitments and responsibilities during the implementation and operation of a multiparticipant GIS project. The major concerns addressed in most multiparticipant project master agreements are as follows:

- Identify participants
- Define organizational structure
- Define cooperation
- Define participants' responsibilities/rights
- Define data-sharing specifics

- Define security and confidentiality
- Define cost allocation share
- Define financial and administrative details
- Define prospective partners' or dropouts' role/rights
- Identify who keeps "master" GIS database
- Define sale of data issues
- Stipulate that technical issues reside at committee level
- Define maintenance procedure
- Define data quality standards

Any number of additional issues may also be addressed in the master agreement. Each multiparticipant project will have its special requirements. It is important to assure that the issues are addressed in contractual form and that provisions exist to accommodate future changes.

9.5 Funding

There actually may be two levels of project participation: (1) organizations participating in a feasibility study to determine the multiparticipant project's technical and economic viability and (2) organizations contractually and financially committed to subsequent GIS database implementation participation.

Once the database issues are defined, funding options can be addressed. Several interesting and viable approaches to multiparticipant project funding have been used. These approaches include public bond issues, private financing, the levying of user fees, special assessments, multiyear participant project financing agreements, private corporation contributions, and integration with already funded federal, state, or provincial programs. For some multiparticipant projects, data may be marketed and sold to private sector entities such as market research firms, real estate/development firms, insurance companies, and banks.

Participants purchase several items when they opt for the multiparticipant approach to GIS. First of all, they are buying a land base that combines a variety of data. All participants weigh the relative value they place on the database, negotiating an equitable cost allocation.

Second, some participants may purchase additional land base or facilities data, tailoring the GIS database to their specific needs. This activity may create a difficult land base maintenance environment and has to be taken into account during the discussions of database maintenance.

Third, participants may purchase land base maintenance, or they may decide to create it themselves. The purchase of maintenance is an item with quantifiable costs for all parties based on a predetermined maintenance approach, sources, responsible entities, frequency, data distribution, and so on. Once costs are firmly estimated, various algorithms can be developed to determine their equitable allocation.

Multiparticipant projects vary considerably in makeup, requirements, and funding approaches, but two key factors drive the equitable allocation process: (1) a balance between the cost of the data to a participant on one side and that participant's minimum acceptable functional data requirements on the other and (2) the value placed by the participant on the access to that data.

Ultimately, the cost allocation is negotiated, and there is no set formula that governs the negotiation process. The parties reach a middle ground, based on a realistic assessment of the economic value of the database to each party and on the common good of a multiparticipant land base project going forward.

A common misconception at the outset of this process is that every party places the same value on the land base. In fact, the land base usually holds a widely disparate value to individual participants, depending on various factors, such as accuracy. For example, a city public works department typically requires a greater degree of accuracy in its land base than an electric utility. The city's minimal scale requirements may be on the order of ±2-foot accuracy, while the telephone company may need only ±20- to ±40-foot accuracy, and gas or electric utilities may need accuracy in the ±5- to ±10-foot range.

In this case, while the electric utility does not require the city's accuracy level, it does recognize the value of increased accuracy to its own operations and knows that it cannot, realistically, justify the added cost of building this database alone in the near future. Also, by sharing a portion of the land base cost, the utility receives a higher accuracy level for less cost than a lower accuracy land base built alone. On the other hand, the city requires the accuracy but may not be able to afford it. It can reduce the increased cost by spreading part of the increase to other participants. On this basis, the parties can begin to negotiate an equitable allocation.

Multiparticipant projects are funded internally by each participant in a variety of ways as follows:

- Labor productivity benefit offsets
- Asset-related cost reduction and cost transfers
- Small surcharges on various permits

- Funds generated from improved utility and tax bill collection
- Increased product and service sales
- Funds generated through improved permit tracking
- Funds generated through improved marketing
- Cost reduction through database integration with GIS database

There will always be significant pros and cons associated with a multiparticipant project. For example, if one of the participants withdraws from the group or cannot meet its annual funding obligations, then the project may suffer. On the other hand, if one or more participants show leadership, they can help pull the other participants up to their level and help drive the project to a successful completion. However, due to its economic advantages, the multiparticipant approach is usually preferred over single participant projects that accomplish the same goals but with a duplication of effort and expenses.

9.6 Data Ownership

The multiparticipant project may, by its nature, require ownership and operation of a common data center and may include either conversion facilities or land maintenance and data distribution facilities. If a common data center is used, asset ownership should be clearly specified. If multiple-year payments are required, these payments should be fully scheduled for each participant.

Nondata asset ownership issues are easier to resolve than data ownership and maintenance issues. Data ownership and maintenance issues, whether administered through a vendor or other third party, must be clearly spelled out. Typically, one participant will take the lead in these responsibilities.

9.7 Liability

It is important to recognize that GIS data are an asset to the multiparticipant organization. If the GIS database can be perceived as an asset, then it becomes similar to any other asset owned by a public organization.

One overriding consideration must drive the development of the shared GIS database: it is developed to foster user access. The liabilities that this access may entail must be defined and dealt with proactively. One difficulty to overcome is that many will expect a report or record generated by computer to be accurate just because it came from a computer.

Some say that the accuracy of a GIS database is only as good as the data used to create it. Provisions must be made to establish nonperformance liability among the various participants. When information is sold from the GIS database, a disclaimer as to the accuracy of the source data is necessary in order to protect the organization supplying the data.

Another major issue relates to copyrighting databases. A copyright can protect the multiparticipant organization's investment in the database. It assures that the creators of the database retain the sole right to its use. Conversely, any external copyright protection that exists on the database will have to be studied to see if it prevents database distribution among project participants.

A licensing mechanism can establish a contractual relationship between the multiparticipant organization and the purchaser of the data. Contract law is more directly enforceable than copyright law. The license can specifically address the conditions under which data can be used. In addition, the license contract can address liability and any other issues related to the use of the data. A summary of action items for research on liability issues are as follows:

- Obtain copies of state Public Records Act
- Obtain copies of state Freedom of Information Act
- Identify copyright laws (state and federal)
- Determine definition of public records and access issues
- Identify map maintenance liabilities
- Identify state map information distribution programs
- Identify state statutes on:
    - Public information costs
    - Manual map product costs
    - Data resale
    - Digital data security issues

Another issue that needs close attention is the interpretation and application of open access laws that vary from state to state. It is absolutely imperative that specialized legal counsel be retained to help research and identify pertinent state legal issues impacting public agency responsibility to provide open access to public GIS information. Sometimes guidelines for pricing information provided to the public are set by law. A study of open access laws should be done months in advance of the development of the draft agreements.

Multiparticipant project contractual and legal issues are becoming more complex as projects grow in scope and in the number of participants. It is important to acknowledge that projects usually take years to fully implement. Often, the original players involved in the project will not be in the same positions when the project is fully operational. Project team members should attempt to develop contractual approaches that assure the GIS database is used to the optimum extent, by both project participants and the public at large, while limiting database access that may increase liability to unacceptable levels.

Thoughtful development of the written agreements controlling multiparticipant GIS projects will provide a solid foundation for their continued growth and health. Change occurs within and happens to each participant. Having a formal written agreement of the intents and responsibilities of the participants is not only good business, but may provide the long-term guidance necessary to keep a successful project on track.

9.8 Risk

GIS multiparticipant projects are not without risk. Things can and do go wrong, but most negative situations can be avoided by tuning into the early warning signs of trouble. Common project traps are listed below; these risks are also inherent in single-participant projects.

1. Insufficient project commitment by management, such as understaffing, underfunding, and weak political support.

2. Lack of vision by executive and technical committee members.

3. Inadequate levels of technical expertise committed to the project.

4. Attempts to stretch out the project length in hopes of reducing short-term funding needs.

5. Participants taking too casual an approach to the project. These projects must have a sense of urgency to be accomplished in a reasonable time frame.

6. Failure to come to terms with who should maintain data and to achieve the necessary commitment by the selected party. From the moment data is converted, it must be kept current.

7. Reluctance to spend time and money developing a conceptual definition of the project and preparing an implementation plan.

8. Attempting too much too soon, such as demanding too high a database accuracy or asking for a large data conversion project to be completed in too short a time. Projects must be completed using an orderly and planned approach.

9. Failure to educate executive committee members so that they too can benefit from a reasonable learning curve.
10. Ineffective communication channels that impede or slow important information exchange.
11. Hidden agendas within one or more organizations.
12. The inability to reach a consensus within one organization or among organizations.
13. Unclearly defined QA standards.
14. Data not maintained to original standards.
15. Failure to thoroughly measure (quantify) actual benefits achieved.

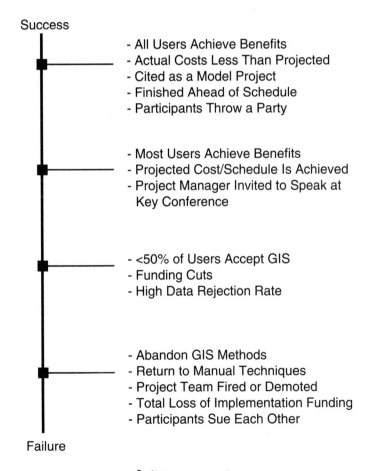

Project success spectrum.

9.9 Feasibility Study

The technical requirements for and the savings associated with implementing a common GIS land base are established through a multiparticipant feasibility study. To perform this study, an agreement called a memorandum of understanding is used to tie parties in a spirit of cooperation to jointly study the issue (and usually commit minor funding for such study), but it does not bind them together to build the common GIS. Once the feasibility study is complete and the requirements and benefits have been identified, a more detailed agreement is necessary to specifically define all aspects of the implementation process.

9.10 Database Contents

One of the principal tasks of the feasibility study is to establish the actual content of a database to be shared. Invariably, different organizations have different data needs, different ways of representing things, and different information needs. The following is a sampling of the questions that have to be answered:

1. Specifically, what geographic and facility data features are of interest to most or all participants?
2. Who will provide what data sources (maps, records, existing files)?
3. What is each of the participants willing or able to purchase?
4. What GIS products will have to share data?
5. What database design will satisfy the requirements of all participants?

These questions are not trivial and can be the subject of considerable debate. Therefore, the answers to them must be thoroughly documented and signed by all participants.

9.11 Maintenance

Data maintenance is a key issue. Some participants may not be at all interested in performing maintenance on the shared database, while others may require having maintenance access to it. If a shared database is edited at different rates by participants, it may become impossible to maintain the database in a central fashion. There is usually no simple solution to this because participants doing their own maintenance may not want to pay for centralized maintenance, while others request the centralized approach. The easiest solution, if acceptable to all, is the sharing of a centrally maintained land base with all participants maintaining their own database layers. Some of the questions to address include the following:

1. Which participants have specific, unquestioned responsibility for maintenance, information currency, and update distribution?
2. Will this responsibility be for all data or for just a subset of the database?
3. Are multiparticipants responsible for maintaining different portions of the database?
4. What happens if the responsible party abandons maintenance and distribution responsibilities at some future point?
5. Do other participants have recourse to enforce maintenance?

These questions have a significant impact on every participant. Details concerning data maintenance responsibility should be clearly spelled out in the initial multiparticipant agreement. If possible, the feasibility study should deal with these issues as early as possible.

Multiparticipant GIS studies and projects have become significant trends. About three quarters of public sector projects have some multiparticipation, and this represents only the tip of the iceberg of expected future data sharing for GIS in the 1990s.

This approach does provide implementation complexities for data conversion contractors, legal and purchasing departments of participants, consultants, and vendors. Economic savings can be considerable, however; and the opportunity to improve products through automated data exchange is very attractive.

Chapter 10

Outsourcing GIS Data Conversion

10.1 Introduction

To understand the role of the GIS data conversion contractor and the evolution of the different types of conversion contractors operating in the GIS marketplace today, it is first necessary to grasp the broad range of tasks that fall under the heading of GIS data conversion. A brief discussion of the evolution of the GIS data conversion industry will show that the categorization of 1990s conversion contractors is largely based upon events of the 1970s and 1980s. The chapter concludes with coverage of the issues involved in selecting a conversion contractor and a discussion of trends in the GIS conversion contractor arena.

Evolution

The GIS data conversion industry has undergone an interesting evolution since its inception. The early days of GIS actually predate the coining of the GIS and AM/FM acronyms. Private firms and government agencies were involved in the early projects cited in chapter 1, "AM/FM/GIS and Their Markets." The introduction and general acceptance of the now familiar acronyms came many years later. Without fully realizing it at the time, these first digitizing contracting firms were the true pioneers of the GIS data conversion industry. In fact, many of the individuals involved in early GIS data conversion activity used the expertise they gained to become industry leaders in consulting, system vending, or data conversion.

In the earliest days, the firms doing the conversion work for GIS projects were, in essence, extensions of the utility or government agency sponsoring a project. The contractors bought a few minicomputer-based digitizing stations, hired some draftsmen and programmers, leased office space, and began digitizing land and facilities data.

They were extensions of the sponsoring organization in that they often had only a single contract for their revenue base and clients engaged their services simply to expedite the data entry of maps, drawings, and other sources.

As the concept caught on and computer graphics technology progressed, contracts requiring more sophisticated services began to evolve. During the 1970s, data conversion companies started to see requests for services (e.g., surveying, aerial mapping, and precise calculation) that were beyond their immediate range of offerings. Accordingly, for projects requiring these other services, subcontracts were given to firms specializing in aerial mapping, etc. These initial business relationships between data conversion companies and firms specializing in other mapping-related activities planted seeds of interest, the cultivation of which will be discussed later in this chapter.

During the 1970s, pioneer firms also began to expand the range of services offered. It seemed as if each new data conversion project required at least one new service that had not been offered on previous projects.

Additional record and map preparation (scrub) and compilation personnel were added, as the importance of having source documents ready for digitizing or keyboard entry was fully realized. Stated another way, the economic benefits of having scrub activities conducted at an inexpensive drafting table as opposed to an expensive minicomputer-based digitizing station, were realized about this time. The term scrub, as applied to source records and with its obvious cleansing connotation, began to appear in conversion contractor marketing literature as a key service offered.

Field inventory services were either developed in-house by the larger and more successful conversion contractors of the 1970s or were subcontracted out to other firms. Interestingly, the prevalence of full field inventory for AM/FM projects has diminished since the 1970s, although the data quality advantages remain just as valid in the 1990s. Partial field inventories are still quite common and very important. In the 1980s, an increasing number of firms became interested in supplying data conversion services. The list of fully qualified conversion contractors grew from some dozen in 1980 to well over several hundred by the early 1990s. The 1980s is thought of as the "decade of experimentation" in the conversion contracting industry. By 1980, there had emerged a viewpoint (justified or not) in organizations that the price of conversion was just too high. This was in spite of the fact that several major conversion contractors had failed during this period and those that were surviving were doing so primarily through their generous and committed corporate ownership, which continued to pour operating and development dollars into some firms that were losing money on very competitive data conversion contracts.

Rather than focusing on improving poor quality source documents, which next to high positional accuracy is the most significant contributor to high conversion costs, conversion contractors began to look for technology-based solutions to reduce costs. At a minimum, this made it appear to their client community that technological solutions were able to offset some of the poor recordkeeping of the past. Conversion contractors yielded to pressures to reduce prices by experimenting with concepts such as off-line text entry, voice data entry, optical character recognition systems, large format document scanners, and so on. These solutions are valuable, of course, but they cannot entirely offset the cost associated with poor quality source documents.

Services

As a result of the industry's 20-year evolution, a broad range of services are offered by today's GIS data conversion contractors. Although not all services listed below are offered by each conversion contractor, the industry as a whole can provide these data conversion services:

- Physical GIS database design and implementation
- Deed research
- Records compilation
- Records scrub
- Land records digitizing
- Records posting
- Facilities records digitizing
- Conventional ground surveying
- Global Positioning System surveying
- Aerial photography
- Analytical triangulation
- Remote sensing
- Image interpretation
- Application programming
- General GIS software development
- Field inventory
- Scanning
- Quality assurance and quality control services
- Data validation services
- Data translation/migration

External vs. Internal Conversion

The extent to which an organization utilizes the range of services available from conversion contractors depends upon a basic decision. The organization must first determine what portions of the overall data conversion effort are to be conducted externally, and which are to be conducted internally (in-house).

In-house conversion.

Some aspects of this decision process are very straightforward. For example, few (if any) utilities and municipalities own an airplane equipped for precision aerial photography. Accordingly, if a project requires aerial photography, it is safe to assume that such services will have to be contracted.

On the other hand, services such as ground surveying, records scrub, records posting, and even GIS database design could be undertaken internally. As a general rule, most data conversion-related tasks associated with a major GIS implementation are contracted out. Decisions concerning the specific tasks to be performed by the conversion contractor involve careful and detailed analysis of cost, schedule, resource availability, and other factors.

A partial list of considerations addressed during such an evaluation includes the following:

- Level of effort required for conversion task
- Equipment availability
- Personnel requirements (during and after conversion)
- Anticipated internal conversion production rates
- Internal versus external conversion schedules
- Internal versus external hourly rates (costs)
- Conversion learning curve versus maintenance learning curve
- Complexity of source documents
- Confidence in contractor's ability to interpret sources
- Version of software being installed
- Precedence of conflicting activities
- Internal availability of specific GIS skill sets

10.2 Relative Market Presence

Several of the largest and most successful firms have achieved their size (in terms of number of employees, number of workstations, etc.) by acquiring or merging with other firms that had some previous experience in GIS data conversion or GIS data conversion support activities. The most successful conversion contractors in the business today seem to have no common profile in terms of the range of services provided. One contractor attributes its success to the fact that it is a full-service organization. The firm owns its own aircraft, cameras, surveying equipment, GPS equipment, photo processing lab(s), workstations, digitizing tablets, etc., enabling it to offer the full range of data conversion services. Another successful firm specializes in field inventory and conventional digitizing services while subcontracting all ground surveying and photogrammetric services.

If there is a common trait among these diverse firms, it is their ability to seek and win data conversion work in the highly competitive GIS arena. This is a result of effective marketing coupled with sound project experience, as attested to by a good reputation in the industry.

10.3 GIS Data Conversion (Focus) Contractors

The first category of GIS data conversion contractor to be explored is the conventional contractor. This term is applied to some of the early data conversion contractors who have steadily grown in terms of personnel and digitizing equipment quantities, but did not have their roots in photogrammetry, surveying, or engineering and have not expanded to include such services. However, to handle some very specialized markets (e.g., independent telephone companies), there still exists a handful of successful (usually relatively small) firms that have concentrated their marketing, development, and production efforts almost exclusively on map digitizing and records conversion. The characteristics usually associated with this type of conversion contractor are as follows:

- No aircraft
- No photogrammetric equipment
- No photo lab
- Large number of digitizing workstations
- Often second-hand workstations and peripherals
- Low-cost, unskilled labor for entry-level positions
- Broad range of CAD and GIS software
- Product-oriented, not "leading edge" in terms of technology

For the most part, firms fitting this profile view the GIS industry from the map digitizing perspective. That is, the majority of their production and development efforts entail the digitizing of clients' existing maps, drawings, etc. They generally possess many digitizing workstations, plotters, and drafting tables to support records preparation, digitization, and other tasks. Some possess digital scanners. In instances where a client requires ground survey, photogrammetry, or other services that the conversion contractor does not offer, the contractor will subcontract to a suitable firm. Because of the increased complexity of typical GIS conversion contracts today, the majority involve one or more subcontractors.

10.4 Aerial Mapping Firms

Aerial mapping firms are the next type of GIS data conversion contractor. These firms saw that they had the opportunity to provide a wider range of GIS data conversion services, and not just the photogrammetric component. The characteristics usually associated with aerial mapping firms that have expanded into GIS conversion activities follow:

- Possession of aircraft or long-standing relationships with aerial photography firms
- Strong photogrammetry background
- Possession of stereoplotters and standard workstations
- Often a civil engineering/surveying perspective
- Some degree of in-house surveying capability
- Lesser interest in conversion contracts that lack a photogrammetric component
- Higher pay scale than straight conversion shops
- Generally an in-house photo processing lab in-house

The use of photogrammetry goes back to the early to mid-1970s, when the first GIS conversion contractors began utilizing aerial photography firms and photogrammetry firms as subcontractors. In those days, the deliverable products to the conversion contractor were often nothing more than contact prints of the aerial photographs. Subsequently, the conversion contractor would digitize the streets, railroads, and other visible features required for the GIS from these photo prints.

Once the aerial photography firms realized that there was additional revenue to be earned by expanding their role in the overall process, they began to purchase CAD or GIS of their own. By 1980, the common deliverable from a photogrammetric firm to a conversion contractor was not a map, but, instead, a digital file of the land base data. The conversion contractor would then add (digitize) the facilities information to the land data and perform the necessary validations on both the land and facilities data.

Over time, these photogrammetric firms continued to expand their role in the GIS data conversion process. Many obtained contracts that involved relatively simple facilities networks such as storm sewer systems. This provided the basis for even further progress into the conversion of other data such as for water and sanitary sewer systems.

From an overall and neutral industry standpoint, one weakness of this type of firm is a sometimes narrow focus on GIS data conversion approaches. Because of their photogrammetric heritage, many of these firms seem to believe that all conversion projects require a photogrammetric solution. The inflexibility that results manifests itself in recommendations to clients in terms of the following:

- Positional accuracy requirements
- Input scale requirements
- Output scale requirements
- Database content
- Technological options

10.5 Engineering Firms

The next group of GIS data conversion contractors is composed of engineering and surveying firms. This group also joined the GIS conversion market through subcontracting. The characteristics usually associated with engineering firms that have expanded into GIS conversion activities as follows:

- Try to offer their conversion services on a professional services basis
- Have a major civil engineering/surveying division or business unit within the organization
- May possess stereoplotters
- May possess standard workstations
- Maintain a higher pay scale than straight conversion shops
- May run an in-house photo processing lab
- Either possess aircraft or have long-standing relationships with aerial photography firms
- Usually have extensive surveying capabilities

These firms' interest in data conversion goes back to the early to mid-1970s, when the first conversion shops began utilizing engineering or surveying firms as subcontractors. However, the pull-through effect in this area was not as pronounced as it was for aerial mapping firms. Engineering firms had their own avenue into this market. Many had previously done significant amounts of work for utility, telephone, municipal, and government organizations that were about to implement GIS. Previous contracts included land surveying, highway design, bridge design and inspection, drainage projects, pipeline projects, and manual map drafting and maintenance. Many of these activities involved the use of CAD systems. Some vendors of CAD systems also offered mapping modules as related products. As a result, GIS became a natural extension of the services offered by these firms.

Because of the pre-GIS work these engineering firms performed, they today have a perspective that distinguishes them from other types of conversion contractors. For example, in their pre-GIS work with local governments, engineering firms usually dealt with the public works departments. This instilled a definite work orientation toward GIS applications such as pavement management, water system design, and sewer system design. In other words, an orientation toward public works solutions can be noticed. This perspective manifests itself in the manner in which these firms approach the following issues:

- Need for three-dimensional coordinate systems and displays
- Extent of construction detail
- Positional accuracy requirements
- Use of COGO input
- Input scale requirements
- Output scale requirements
- Database content
- Data maintenance workflow

Throughout the 1980s, engineering firms began making inroads into the local government and utility market sectors. However, some were much more successful than others. One national firm ventured into GIS in a very big way. They purchased large quantities of state-of-the-art equipment, recruited top-notch personnel, and landed a much-sought-after major utility contract. They struggled in the industry for several years and then got completely out of data conversion due to financial losses.

Other smaller firms did little more than mention to their existing clients that they had recently added GIS capabilities. Several of these firms are now major players in the GIS data conversion industry.

10.6 Automated Conversion-Oriented Contractors

The newest type of GIS data conversion contractor is composed of those firms that emphasize the broad use of digital scanners in their approach to conversion. To understand the distinction being made by this type of conversion contractor, one must realize that virtually all other conversion contractors possess and use scanners to some extent. Also, most successful firms in the automated conversion contractor category got their start in some other field and later began to heavily apply scanning to GIS conversion projects.

Under certain circumstances, scanning GIS source documents offers significant cost savings as compared to conventional board digitizing techniques or stereocompilation. In addition, scanning can provide an edge in source document handling, since the contractor can scan a source document quickly, even at the client's office, requiring access to the document for scanning for less than an hour.

The level of automation that can be achieved via scanning varies among individual contractors. These differences are attributed to the specific scanning or raster processing system being used.

The more sophisticated scanning systems offer (or at least claim to offer) real-time vectorization, batch text recognition, batch symbol and pattern recognition, and automatic text-graphic relationships. Other systems offer fewer of these capabilities. Within the GIS arena, the greatest challenges facing this type of conversion contractor are the poor quality of source documents used for input and the large number of source types (each of different size, scale, and legibility) involved in a typical data conversion project.

Regardless of the approach taken and the hardware or software used, the amount of document preparation effort to conduct scrub, posting, and other tasks is not reduced by a scanning approach. Accordingly, on bids that involve significant amounts of records preparation prior to data conversion, automated conversion contractor bids are usually equal to or higher than bids from conventional conversion contractors.

Most automated conversion contractors approach all data conversion projects with the scanner as the core of their process. The same criticisms raised against aerial mapping firms for their occasional procedural bias are levied against automated conversion contractors. That is, critics complain that automated conversion contractors apply scanning technology even in cases where it may not be appropriate. Because of continued fascination with scanning, contractors of this type will likely grow in number.

10.7 GIS Vendors

Some GIS vendors engage in data conversion activities. Historically, this has been a very limited endeavor, restricted to data conversion in support of benchmark preparation, pilot project areas, etc.

Vendors of GIS can be successful in obtaining data conversion work by contractually coupling the data conversion tasks to the sale of the GIS or to related GIS support services. In fact, the GIS vendor community is clearly divided into two groups: those who view data conversion as part of a comprehensive GIS package, and those who prefer to avoid the complications and risks inherent in the data conversion business.

10.8 Criteria for Selecting a GIS Conversion Contractor

Several criteria drive the selection of a conversion contractor for that part of a GIS implementation.

External data conversion

Technical Considerations

The first consideration concerns the contractor's technical ability to accomplish the particular data conversion project. To excel in the technical portion of the evaluation, the prospective conversion contractor must convince the evaluation team that the conversion company possesses the necessary background to do the job, and must also present a thorough understanding of the client's unique GIS project needs.

Company History

The next background item is the conversion contractor's history or origin. A firm that has a long-standing history of providing data conversion services will be evaluated slightly more favorably than a firm that has only recently started to offer them.

Full-Service Company

The range of services offered by the conversion contractor is compared to the specific services required for the client's project. This topic leads directly to an evaluation of subcontractors in many cases, since about 90 percent of the conversion contractors competing in the market today are not full-service firms. Sometimes, though not always, a slight advantage goes to the truly integrated, full-service firms that will not require the use of any subcontractors because the company possesses its own GPS equipment and crews, aircraft, stereodigitizing equipment, digitizing workstations, and so on.

Company Location

The location of the conversion contractor's primary conversion facility is another consideration. With overnight delivery and fax machines, this is of considerably less concern than it was a few years ago. The majority of conversion contractors operate one or two centralized data conversion facilities. These firms have demonstrated the ability to manage projects that are geographically remote from their data conversion offices. For cost and logistical reasons, local conversion contractors sometimes score somewhat higher during the evaluation of this item. More commonly, the conversion contractors shortlisted to receive the conversion RFP are hundreds or thousands of miles from the client and can compete very effectively.

Organizational Structure

The conversion contractor's organizational structure is also carefully evaluated. There exists, for the most part, a similarity among the personnel organization of the successful conversion contractors. Most utilize a classic hierarchical structure. Evaluation teams do tend to penalize firms that have organizational aberrations. For example, a conversion contractor that has quality assurance personnel reporting to a financial officer would not be evaluated favorably.

Corporate Ownership

Corporate ownership is another background factor. Given the number of conversion contractors who have failed commercially, this item has been given steadily increasing importance in the overall evaluation process. Evaluation team members give the highest scores in this category to firms with strong backing and strategic power.

Other aspects of corporate ownership involve other corporate ties or parent corporation orientation or strategic complementary areas to GIS conversion. This provides the potential for leveraging other aspects of related fields including technology transfer and price-related transfers.

Company Revenue Base

Closely related to corporate ownership is the conversion contractor's revenue base over the previous three to four years. Both the corporate revenue picture and the specific GIS revenues are examined. Conversion contractor evaluation teams tend to penalize, and in many cases disqualify, firms with less than $2 million of annual GIS revenue. The more successful, lower risk conversion contractors now generate between $5 million and $20 million of GIS revenue annually.

Another financial area evaluated is previous project size. Truly successful firms have demonstrated the ability to successfully manage multiyear, multimillion-dollar data conversion projects. Successful conversion contractors also have not incurred or have successfully mitigated or eliminated litigation.

Company GIS Commitment

The conversion contractor's commitment to the GIS industry is also frequently evaluated. The successful firms are those that have taken active roles in GIS industry-related professional organizations, such as the Urban and Regional Information Systems Association (URISA) and AM/FM International, by frequently presenting papers, serving on committees, and offering financial support. Participation in GIS vendor user groups is also positively scored.

Company Resources

The resources that a conversion contractor can bring to bear on a data conversion project are a critical consideration during the evaluation process. Naturally, the required level of resources is a function of the scope of the particular project. A conversion contractor possessing 10 to 20 workstations would probably be considered a legitimate contender for only a limited project. Such a firm would not be considered viable for a large utility data conversion project with a tight schedule.

Personnel Experience

In addition to scrutinizing conversion contractor experience, evaluation teams also look at the resumes of key personnel proposed to work on the project. Generally speaking, experience counts more than education. The most highly valued experience is that which is most directly applicable and relevant to the client's data conversion needs. The most successful conversion contractors are those who have developed and/or recruited project managers, software engineers, photogrammetrists, digitizers, compilation technicians, and quality assurance personnel with more than 5 (preferably 8 or 10) years of experience on GIS conversion projects. Few can boast personnel in those positions with 15 or more years of experience. Of course, the availability of the personnel cited in the proposal must also be examined (such as parallel commitments on other projects, or other conflicts).

Company Experience

Often, when determining which conversion contractors are to receive the conversion RFP, the single most important factor is whether they have production experience in delivering data to the GIS selected by the client. Evaluation teams typically disqualify conversion contractors that have not had such previous experience. The disqualification sometimes extends to those who have a software license for the target GIS, or who have their first pilot project underway on that system, but who cannot cite project references for the vendor-specific GIS.

Technical Plan of Operation

The technical plan of operation presented by the potential conversion contractor is usually viewed by the evaluation team as one of the most important portions of the conversion contractor's proposal. The most successful conversion contractors are those who adequately address the myriad technical considerations related to the preconversion, conversion, and postconversion phases of the overall project.

For the preconversion phase, successful conversion contractors present detailed procedures concerning their approach to source document gathering and tracking, scrub operations, and so on. Evaluation teams favor conversion contractors that assist the client in understanding precisely which tasks will be conducted by the client and which will be conducted by the contractor.

Predictably, the conversion phase comprises the most voluminous part of the response from the conversion contractor. This section of proposal presents the step-by-step process the conversion contractor will use to convert the data. Through a combination of flowcharts and associated narrative descriptions, the successful conversion contractor will do the following:

1. Acknowledge all source documents and their routing through each process.
2. Describe the hardware and software to be used for each process (often different from the client's GIS).
3. Describe which operations will occur on which equipment (e.g., stereodigitizing, map digitizing, key entry).
4. Describe the compilation effort involved (i.e., extraction of data from multiple source documents).
5. Describe the use of macros and customized data conversion software.
6. Distinguish between manual edits and automated edits.
7. Distinguish between interactive and batch automated edits.
8. Describe the error detection process for all types of edits.
9. Describe the error correction process.
10. Describe integration or use of field inventory or verification work.
11. Describe the final deliverable product(s) generation process (especially the data translation process, if appropriate).
12. Discuss upcoming innovations that may improve any of the above processes.
13. Thoroughly describe the QA/QC methodology.

The postconversion phase includes delivery of the conversion products (plots and magnetic tapes) and the client acceptance and rejection cycle. Successful conversion contractors in this area are those that present a convincing case that they have thoroughly thought through the problems associated with data translation, incremental data deliveries, software revision level compatibility, optimal file sizes, file naming conventions, meeting client acceptance criteria, and error correction and redelivery of database information.

A major factor in evaluating all conversion contractor proposals is a check of references from existing and past clients. Clients place extra weight on recommendations and comments that are provided by their industry peers. The following list contains a representative group of topics addressed during the conversion contractor reference check process:

1. A rating of performance on a scale of one to ten.
2. Ability to meet schedules.
3. Ability to deliver consistently high-quality data.

4. Ability to interpret source documents.
5. Experience and capabilities of conversion contractor project manager.
6. Number of files delivered when in full production.
7. Technical expertise.
8. Problems encountered and how resolved.
9. Name of the project manager for the conversion contract.
10. How easy it was to work with the conversion contractor.
11. How cooperative the conversion contractor was.
12. Any invoicing, payment, or financial problems.
13. Would this conversion contractor be selected for additional work?
14. Other projects that the conversion contractor is doing or has done.

Price

The importance (weighting) of price in the conversion contractor evaluation process varies dramatically among data conversion projects. Although always a consideration, data conversion projects are not necessarily awarded to the low bidder. In some instances, the evaluation team must consider and evaluate solid evidence as to the added value received by awarding the work to a conversion contractor other than the low bidder.

Price remains a very influential factor in the selection of conversion contractors. Pilot project prices can vary widely and are sometimes influenced by marketing strategies (e.g., a desire to break into a particular market sector). This is often acceptable for a pilot project when the client does not have schedule constraints and is willing to take a price risk. The risk is minimized when the low price is presented by a conversion contractor that scores high in technical categories or when the low price is associated with an established conversion contractor with a track record of successful GIS data conversion projects.

When budgets and schedules are critical, there is a tendency to go with the lower perceived risk associated with the technical score winner. This is especially true when the data conversion is beyond a pilot project and is associated with full GIS project implementation.

Data Conversion Cost Factors

Upon performing an analysis of past GIS data conversion projects, it becomes apparent that many factors influence a successful conversion contractor's pricing formula. The concept of volume discounts, common to many types of businesses, also applies to data conversion pricing. The start-up development and training costs associated with a pilot project, for example, can be a significant portion of the total data conversion cost if the award is for the pilot project area only. Alternatively, if the award is for the client's entire service territory, these up-front costs can be distributed across a much broader base. Regardless of how a pilot project is handled, economies of scale in conversion unit costs only occur during the major data conversion phase and are usually inversely proportional to the size of the project.

Number of Contractors

This topic refers to a client's decision to utilize multiple conversion contractors to do the GIS data conversion. Generally speaking, overall project costs are lower with a single conversion contractor, since the start-up costs are only incurred once. Conversely, if multiple contractors are involved, the client will pay for multiple learning curves, whether or not they are cognizant of this fact. The major benefit of multiple contractors is reduced risk to the project schedule.

Target GIS

The target GIS selected by the client is also a factor influencing conversion contractor pricing, although over the past three years this has become less pronounced. For the most part, successful conversion contractors have developed customized conversion software on their platform of choice. They then conduct the majority of the conversion effort on that system, using data translators to produce the client's required format. This is particularly true for the land base (photogrammetric) portion of most data conversion projects, since not all popular GIS support a direct stereo photogrammetric interface.

Data Quality

As expected, the required data quality of the converted GIS database also impacts data conversion costs. Within the realm of GIS, two types of data quality are typically addressed as part of the client's acceptance criteria: graphical and informational. Conversion contractors do not control the data quality levels specified by the client, but they do control the degree of impact the quality specifications have on data conversion pricing.

Database Design

The responsibility for GIS database design, implementation, and conversion specifications development also impacts the conversion price for each project. Since the accounting of these costs can be somewhat controversial (i.e., Are they system costs or data conversion costs?), successful conversion contractors must convince their clients that these front-end costs be considered separately. In fact, some projects have placed all start-up costs, data conversion-related and otherwise, on the system side of the ledger. Obviously, this tends to make the data conversion appear less costly.

Source Document Scrub

The amount of required source document preparation (i.e., scrub) will also influence the data conversion price. Again, some accounting considerations surface, since the scrub may be conducted completely or partially by the client. Scrub can be very involved, particularly if it includes redrawing sources, posting outstanding work orders, reformatting text, adding new data items, etc. Cost estimates for document scrub are often requested by the client so that an analysis can be made as to whether it is best done by the conversion contractor or by the client.

Source Documents

Another data conversion price impact factor is the input side of the data conversion equation; that is, the source documents. The number of source document types to be handled, compiled, and converted by the conversion contractor dramatically impacts the price. All other factors being equal, a project utilizing 20 different types of source maps or documents will have a conversion price total that is several times higher than a project that has only two different types of sources.

The quality of the source documents is also a major factor in determining data conversion pricing. Source traits such as legibility, degree of existing symbology standardization, currentness, and reproducibility are carefully analyzed by the prudent conversion contractor before a price is calculated.

Conversion Deliverable Products

On the output side of the conversion equation, the GIS database content, database structure (i.e., level of required intelligence), and deliverable map scales and format have major price impacts. A complete planimetric, cadastral, and topographic GIS database conversion project may be bid as high as 10 times the cost of a simple (graphics only) planimetric base (on a dollar per square mile basis). Since a 1"=200' (1:2,400) map covers four times the area as a 1"=100' (1:1,200) map of the same physical size, there is a direct correlation between scale assignment and total data conversion project price. Successful conversion contractors expend significant effort on working with the client to establish the optimum map scale assignment within the project territory.

Data conversion costs are influenced by the deliverable products required by the client. Some projects specify that a magnetic tape copy of the GIS database and a single set of electrostatic edit plots be delivered. In these cases, the client internally produces the full range of other graphic output products such as index maps, circuit maps, etc. In other instances, the conversion contractor may be required to produce the full set of map products in final form. In the latter case, the data conversion prices will reflect the additional effort on the part of the conversion contractor. Again, successful conversion contractors have a good historical record from previous, similar projects upon which to base their detailed pricing for these deliverables and to be consistently profitable.

Representation

The requirements for representation of the data also affect data conversion pricing. Factors such as specifying whether a street will be mapped as a standard road width, centerline only, actual edge of pavement, or legal right-of-way enter into the pricing picture even for a simple planimetric land base.

Labor Costs

Labor costs are, of course, a major data conversion cost component. Domestic conversion contractors pay the majority of their production workers $5 to $11 per hour. Offshore production workers are paid considerably less. However, most experiments with foreign labor have not produced lower data conversion costs to such an extent that all conversion contractors will follow suit. Nevertheless, those who have effectively implemented offshore conversion shops claim cost reductions of 30 percent while maintaining the same quality standards.

Quality Assurance

Related to the previous discussion of positional, qualitative, and quantitative accuracy is the subject of the conversion contractor's quality assurance methods. In a case where the client has specified ±2 feet positional accuracy and 99 percent informational quality, the conversion contractor must apply very stringent quality assurance measures to ensure that the data will be accepted. On the other hand, the client will have to put in place equally stringent checks to verify compliance with those specifications.

To minimize rework, successful conversion contractors have implemented streamlined quality assurance practices. Generally, all of the converted data will be subjected to project-specific edit software, both interactive and batch. The manual edits are often not conducted on all maps. Manual edit procedures are usually based on statistically sound random sampling techniques. Insofar as is practical, the conversion contractors attempt to precisely duplicate the checks that will be applied by the client during the acceptance process.

Postconversion Processing

Special postconversion processing requirements will also impact data conversion costs. Adjusting coordinates from one datum to another, producing attribute listings, generating special circuit trace reports, and integrating client-provided digital data are examples of additional data conversion costs.

Project Schedule

One of the most significant data conversion price impactors is the data conversion project schedule. This becomes a complicated issue, since the reaction of a conversion contractor to a project schedule, as presented in the RFP, is a function of the conversion contractor's base loading and subsequent available capacity during that time frame. In other words, a short schedule does not necessarily equate to high prices or a long schedule to low prices. Successful conversion contractors, in terms of pricing, are those who have the foresight and good fortune to be winding down one major data conversion project simultaneously with the start-up of another.

Competitive Bids

Another significant factor impacting data conversion pricing is the entire competitive bid process. Successful conversion contractors maintain detailed databases concerning their own and their competitors' bids on previous jobs. The level of a contractor's bid may vary dramatically between two seemingly similar jobs, based on the contractor's backlog of work at the time, the shortlisted competitors for that particular job, the mix of experienced and trainee-level personnel to be assigned to the job, and other factors.

General Bidding Aspects

Exceptions to a client's data conversion specification generally do not negatively impact the evaluation unless the exceptions taken are significant and perceived as defense of a weakness or unless the exceptions are unaccompanied by a full explanation and a suggestion for an alternate requirement. Proposed alternate requirements must not cause additional work for a client unless it is offset by a commensurate conversion contractor price reduction.

Database Complexity

The complexity of a client's database design impacts a conversion contractor's procedures and, therefore, pricing. Complex data relationships and extensive attribute data requirements demand additional effort for data capture, conversion, and quality assurance. This additional effort is reflected in a higher price.

Potential for Other Services

Data conversion pricing may be influenced by the potential for the conversion contractor to provide other GIS services including training, application software development, and ongoing GIS database maintenance. Pricing may also be influenced if the conversion contractor is granted marketing rights to the client's digital land base or contractor-developed applications software, or both.

Proximity to the Client

Although minor, proximity of the client to the conversion contractor can influence pricing. Proximity can enhance communications and occasionally reduce delivery charges. Quicker client response to conversion contractor questions regarding source document information or discrepancies translates into a smoother workflow for the contractor.

Schedule Flexibility

Schedule flexibility can be valuable to a conversion contractor and can result in a discount for the client. A major challenge for conversion contractors is work-load leveling. Flexible client schedules that allow for significantly reducing or increasing the level of work performed should result in a discounted price for this work. Such flexibility allows the conversion contractor an option to avoid overtime labor rates. This flexibility also eliminates the potential need to hire temporary personnel and helps to avoid personnel layoffs as other project work winds down.

Shifts

Conversion contractors occasionally propose the use of a client's GIS during evening and night shifts. This option can reduce costs by allowing the conversion contractor to avoid or postpone a capital purchase; however, some of the projected reduced costs may be offset by increased supervisory, administrative, maintenance, and depreciation costs.

10.9 Risk

The following paragraphs present the common risk factors reviewed during the conversion contractor evaluation process.

Financial Report

In anticipation of the risk assessment concerning possible business failure, many contractors will cite the financial standing of the parent organization (if any). Others include audited financial statements or annual reports to bolster confidence in their financial well-being. Successful conversion contractors are keenly aware of concern in this area because of the number of failures, buy-outs, takeovers, and mergers in the GIS data conversion industry over the past 10 years. They tend to devote substantial effort to convincing potential clients of their stability and the validity of their plans for continued longevity.

Lack of Experience

Another common risk associated with conversion contractors is lack of experience on the target GIS. For conversion contractors in this situation, the response is generally a discussion of their success in developing data translators to or from other systems. Some state that they will acquire the target system if awarded the data conversion contract. Others espouse their staff members' experience obtained on the target system or from previous employment. The level of experience with the target GIS will be apparent in the technical response sections of the proposal.

Loss of Key Personnel

A successful conversion contractor avoids being scored high-risk in this area by including resumes of key personnel with the longest periods of service to the organization. Some will also include a brief discussion of their career advancement policies and professional development activities.

Data Quality Problems

The risk of data quality problems is mitigated by a successful conversion contractor in two ways. First, their proposal presents considerable detail concerning the quality assurance checks that are in place throughout the data conversion processes. Second, the contractor provides client references on those accounts that have had the highest levels of data acceptance. Frequently, this is supplemented by testimonial letters from other recent satisfied clients.

Data Translation Experience

Data translation or migration is another specific risk item examined by a client's evaluation team. Successful conversion contractors provide detailed flowcharts and write-ups depicting the manner in which translation or migration has been performed on other GIS projects. This, too, is supplemented with references from accounts where the translation approach has proven successful.

Poor Project Management

Poor project management is always identified as a data conversion project risk. Predictably, successful conversion contractors propose project managers who have a proven track record and who have a good reputation in the industry. An unfortunate fact of life in the GIS business is that the project manager assigned to the project once it is under way is not always the same individual who was initially proposed. Any conversion contractor's tendency toward this sort of "bait and switch" activity is closely scrutinized in the evaluation process.

Inadequate Data Conversion Software

Inadequate GIS data conversion software is another conversion project risk. This consideration has a very high impact factor since the consequences could be devastating to a GIS project, budget, and schedule. However, this is no longer an issue for the larger conversion contractors. They have developed extensive libraries of data entry, data manipulation, and data editing software. Moreover the software has been used to convert literally millions of records for other GIS data conversion projects. Therefore, the probability of this risk actually materializing is deemed to be very low for established conversion contractors.

Inadequate Workstation Capacity

Inadequate workstation capacity is a high impact risk factor, different from some of the previous ones in that it is often of greatest probability for the most successful conversion contractors. This item is always evaluated by considering the other data conversion projects that a contractor will be performing at the same time as the client's proposed work and the volume of new work likely to be awarded to that contractor during the same period. Conversion contractors rarely turn down work because they are too busy. Even when they are booked, they will continue to bid jobs. Naturally, the highest priced bids are submitted during a conversion contractor's busy periods, and the bid prices tend to get lower as the contractor's work load begins to diminish.

Failure to Meet Schedule

Failure to meet the project schedule is always evaluated as a serious risk. Successful conversion contractors anticipate this fact by citing reasons that force projects to fall behind schedule. Their discussions on missed production schedules focus on factors beyond their control. They often mention the unavailability of source material, delays caused by the GIS vendor's database design or implementation activities, lack of prompt problem resolution by clients, etc. These conversion contractors have learned the importance of public relations as an effective damage control device when schedules are missed. They emphasize the projects that are on schedule and rarely use projects with a history of chronic schedule problems as references.

Staff Acquisition Problems

Staff acquisition problems are sometimes viewed as a risk. Whether or not this is relevant for a particular data conversion project depends on the particular conversion contractor being evaluated, the contractor's existing backlog, and the client's conversion schedule. High risk is associated with cases where new staff will be added by the conversion contractor to accommodate the client's schedule. To determine the probability of this risk, the evaluation team looks at the conversion company's track record of expansion, economic conditions in the conversion company's locale, general work force factors in the area, and the proximity of trade and technical schools with GIS and/or CAD programs.

Market Segment Experience

Market segment experience is another important risk item. An evaluation team for an electric utility, for example, would place a high risk on a contractor that had no previous electric utility data conversion experience. If the proper procurement approach is taken, this item is not an issue. A request for information (RFI) process will typically prevent conversion contractors without relevant market segment experience from receiving the RFP.

Pending Litigation

The frequency, nature, and severity of any recent or pending litigation against the conversion contractor is analyzed during the evaluation period. There have been several recent cases of litigation that directly involve GIS conversion contracts.

10.10 General

There are other conversion contractor evaluation criteria that do not fall under technical, price, or risk categories. These have to do with special situations such as the conversion contractor's willingness to investigate and try innovative solutions, the way client communications are handled, the conversion contractor's general philosophy with respect to GIS, the uniqueness of the data conversion service sought by the client, the fact that the conversion contractor is a minority controlled firm, and the way the conversion contractor handles the bid process. Should the client specify that the data conversion work is targeted as a minority firm set-aside or that other minority considerations apply, then this may become relevant criterion for proposal evaluation.

Several new technologies are being investigated that may result in a considerable savings of money and time during the data conversion process. These involve raster technology and heads-up digitizing, for example. Although use of these technologies might increase the risk for a client, a solution of this sort can be technically and/or economically desirable.

The bid process produces the first mutual commitments between the conversion company and the client with respect to a specific data conversion project. Therefore, it can be instrumental in setting the tone of future negotiations and interactions. If the conversion contractor provides a complete and professional proposal, then the chances are greater that contract negotiations will follow that tone and that the conversion contractor will provide good service. The stage is also set for a businesslike environment in which to conduct additional communications, as needed.

In summary, the major factors in a successful data conversion project are as follows:

1. Attention to cost containment and cost reduction alternatives because of the magnitude of the data conversion project.
2. Meeting schedule requirements (both the conversion contractor and the client).
3. Achieving the prescribed level of data quality.
4. High level of communication between conversion contractor and client project managers.
5. Full-time project managers and sufficient staff.
6. Adequate and timely scrub of source data and quality assurance acceptance by client.
7. Implementation of GIS database maintenance immediately upon acceptance of converted data and subsequent database updates.
8. Converted data quality sufficient to support GIS applications.
9. Users of GIS are satisfied with converted map products, data, and end-user procedures and documentation.
10. Completion of a pilot project to ensure quality and provide for modifications, if needed, prior to full data conversion and implementation.
11. Upper management support of the data conversion project, start to finish.
12. Openness by conversion contractors to site visits in the final selection process.
13. Attention to details including pricing spreadsheet arithmetic.
14. Providing unique conversion methodologies.
15. Exercising innovative and cost-effective techniques to lower costs.

10.11 GIS Data Conversion Trends

Several trends are emerging in the GIS data conversion industry. First, as previously mentioned, the popularity of scanning systems is having an impact. The few conversion contractors who do not possess scanning systems will probably acquire them in the near future. Most will purchase or lease systems, while others may work out subcontract arrangements with firms that specialize in scanning. Already, some evidence of the latter scenario has come up in several bid situations. Document imaging systems and optical disk storage are being viewed as a means to store related information in unintelligent form in support of the GIS.

The most dramatic technological advances will be software driven. Vendors' developments in the areas of workprint generation, work order processing, dynamic segmentation, and interfaces to other systems are beginning to catch up to marketing literature of the past several years. The advent of object-oriented GIS databases will certainly impact data conversion prices as well as GIS applications development.

The GIS industry has not yet fully explored the utilization of large format digital holography, although several futurists within the industry predict this will emerge as a natural technological extension. They believe this technology will be particularly beneficial for engineering applications. Once commercially available and fully integrated with GIS, large format digital holography could provide a realistic three-dimensional model of actual field conditions for designing, planning, and troubleshooting purposes.

Another surfacing phenomenon (it is too early to identify it as a definite trend) is that conversion contractors as a whole are beginning to defend themselves against charges that conversion costs are too high. One effective way they have accomplished this is by intentionally reducing the scope of work on complex projects. That is, the conversion contractors have negotiated with clients to shift more of the records preparation (scrub) burden back to the client. In the case of one large electric utility, the external conversion costs were reduced by 60 percent when the client agreed to perform all records preparation work.

The industry continues to view offshore data conversion with ambivalence. Some organizations are willing to try anything to reduce data conversion costs. Without question, offshore data conversion offers the advantage of lower labor costs. On the negative side, the issues of source document logistics, language barriers, productivity, and quality assurance have yet to be solved to everyone's satisfaction. Some organizations that have utilized offshore conversion contractors are pleased with the results. Others have conversion horror stories to tell. While definitely too early to make any comparison between trends in GIS conversion contracting and those in other segments of the U.S. economy (e.g., manufacturing), domestic conversion contractors are aware and sometimes concerned about the prospect of foreign competition.

To examine the characteristics of the successful GIS data conversion contractor of the future, it is important to understand present conditions. In recent years, there has been a continued shakeout in the GIS data conversion industry. Several conversion contractors have stopped doing business. Other extremely large firms have been created as a result of mergers and acquisitions.

Successful conversion contractors of the 1990s will likely have the following characteristics:

- Proven track record on multiple GIS projects of various sizes and types
- Exceptional technical personnel and management teams
- Experienced production staff
- Adequate number of workstations to meet increasingly aggressive production schedules
- State-of-the-art equipment and software
- Regional offices in locations with desirable labor pools
- Ability to deliver in numerous GIS formats
- Substantial financial backing
- Proven ability to handle multiparticipant projects
- Willingness and ability to engage qualified subcontractors
- Willingness and ability to act as a subcontractor
- Some form of strategic alliances

Chapter 11

Impact of Emerging Technologies

11.1 Introduction

Considering the cost of GIS data conversion for most projects and the large quantity of data that usually have to be converted, continual research, development, and implementation of new GIS data conversion solutions are strongly justified.

The introduction of new technology into data conversion has mostly been in the area of automated methods. Developers, both large and small, have pursued building GIS databases using artificial intelligence, rules-based software, and scanning. While these approaches still hold great promise for the future, short-term expectations are tempered with the large cost associated with developing a standard automated conversion process that is easily customized to the specific data requirements for each project.

The high cost of conversion has forced users to re-evaluate their data needs and to change the data specifications accordingly. This has resulted in changes in conversion methodologies to accommodate raster data capture, incremental vector data capture, and hybrid raster-vector data capture that meet budgetary and scheduling constraints and still provide an appropriate benefit level. Additionally, an increasing need for integration with existing digital systems will continue to impact conversion methodologies.

Because new data requirements have impacted methodologies, process-oriented data conversion strategies have been developed to optimize the series of activities needed to populate today's GIS databases (see fig. 11.1). No longer are system data requirements viewed as graphics with attached attributes, but as a geographic information system requiring a combination of graphics data (raster and vector), a sophisticated database, linkages to other data, and other items that foreshadow distributed GIS processing. Process-oriented data conver-

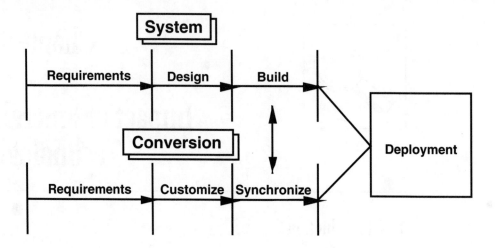

Fig. 11.1. Process-oriented conversion strategy.

sion strategies are developed along with conceptual database and application design, departing from the traditional approach to conversion. The traditional approach has included development of database specifications, selection of a vendor who converts data using a proprietary process, and subsequent data acceptance, but has not provided the ability to completely verify data integrity. The development of a specific process to meet data requirements helps to ensure that data conversion and system requirements are more closely in step.

Process-oriented conversion strategies integrate data acquisition with the business objectives of the GIS. Therefore, data, system, and applications are more closely aligned and begin to provide benefits earlier in the project life cycle. Data are captured as a specific process of the project implementation and are not treated as a commodity provided by a conversion bureau.

Aligned with process-oriented methodologies are new data-driven approaches that are designed along with system requirements. Data-driven approaches provide the ability to integrate, verify, and build GIS databases.

Data-driven approaches place emphasis on creating *enabling* databases as opposed to the more limited *graphics with attributes* type. These enabling databases provide a variety of information system functionalities including business-oriented applications and query-driven graphics.

Data-driven approaches, as well as other emerging technologies, will provide new opportunities for reducing conversion cost and for increasing data quality. Significant emerging conversion technologies are described below.

11.2 Data-Driven Approaches

As many GIS implementations begin to reach maturity, close attention is focused on the realization of planned benefits and the balance of those benefits against the costs. Many large potential GIS implementations have been delayed and even cancelled as a result of the costs associated with data conversion. Even the use of scanning and incremental conversion methods have not always been able to adequately reduce costs.

The data-driven approach is a new alternative to the implementation of GIS technology. The approach seeks to leverage database technology to provide a cost-efficient means of building, maintaining, and obtaining benefits from a GIS.

By focusing on a database approach rather than on CAD-based solutions (traditional automated mapping), a repository of valuable, multiuse data is developed that can enable business, engineering, and locational applications. Figure 11.2 illustrates the value of converting data into an open, nonsystem-specific database. Data may be selected from the GIS database, enabling applications to be integrated with other digital databases or to be managed within an AM/FM/GIS.

The definition and application of GIS data are changing and therefore the approach to conversion is changing. In many cases, source documents have become inadequate to support an organization in today's information-driven business environment. Therefore, many organizations now

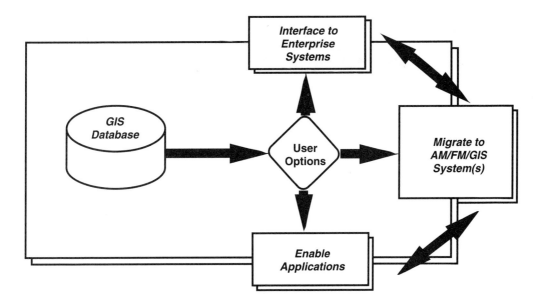

Fig. 11.2. Open nonsystem-specific database utilization.

find benefit in spatial databases containing information such as commercially available addresses, streets, and coordinates. Others relate facilities databases to these commercially available databases through a concept called *geocoding*. Geocoding provides a means for storing spatial coordinates for specific features in a database. Geocoding may also include locating raster images geographically using specific spatial coordinates.

The latest utilization of this approach provides for the creation of graphics on demand through database query (much like a spreadsheet creates pie and bar chart graphics on demand). Graphics generation for GIS purposes from nongraphic land and associated facilities databases is a new area of development that shows significant promise and conversion cost reduction.

The concept of generating graphics from facilities database record information replaces the traditional digitizing process. The data-driven approach is based upon the fact that all *functional* data are contained in the attribute data. If to-from connectivity data were also stored in an attribute field, the system itself could generate the facilities graphics held by points as well as the linear graphics that represent their connectivity. The graphics are generated through a series of rules-based algorithms that are customized during the system development process to ensure that input data will produce output graphics that meet requirements. Figure 11.3 illustrates the geocoding of structures, and figure 11.4 illustrates how graphics are generated from nongraphic databases.

Potentially, the computer-generated graphics may not meet existing requirements for traditional cartographic output maps nor exactly

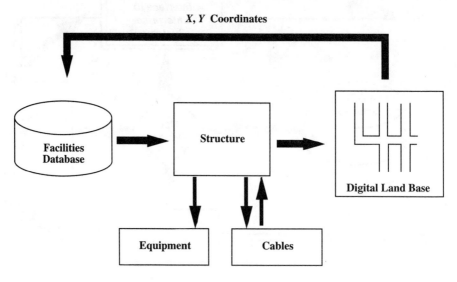

Fig. 11.3. Geocoding of structures.

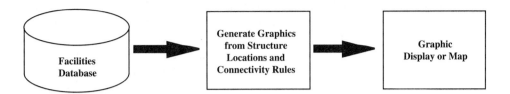

Fig. 11.4. Generating graphics from nongraphic databases.

match the graphic depiction of the source document since the computer determines the optimum graphics view based on the rules-base. The advantage is that, if data are entered without the use of digitizing, conversion costs are reduced drastically.

The other significant advantage is the flexibility of open, nonsystem-specific databases. Data from these systems may be used by multiple GIS environments. For example, spatial analysis may be performed within one set of system applications while engineering applications may be performed within another. This provides more immediate benefits and allows more users to benefit from GIS data. Typically, the maintenance of these databases is similar to the way organizations update other operational system databases.

The database-driven approach to GIS is closely following mainstream database, language, and open system technologies. As a conversion strategy, it provides significant savings and a more powerful data set for users. This approach will influence future GIS implementations even though it will challenge graphic representation requirements.

11.3 Scanning and Related Technologies

Automated Vectorization

Utilization and integration of scanning technologies have captivated the imaginations of data users, systems developers, and data conversion contractors since the mid-1980s. Initially, the vision was to develop a *black box* that would convert drawings into intelligent GIS databases via a completely automated process.

Substantial investments have been made in the research and development of automated conversion systems. In spite of this, the implementation of fully automatic data conversion has not been possible as a cost-effective solution. A variety of factors, such as source document quality, legibility, and reliability, have resulted in additional task requirements that result in costs equal to and sometimes greater than those associated with traditional map digitizing. While text recognition technology continues to develop, hand-drawn, manually maintained source documents have made automated text a near impossibility. Additionally, the creation of sophisticated databases that enable applications is extremely difficult because feature relationships (i.e., conductor splices, lot/text information, and such) must be determined by spatial inference or by massive sets of rules. Figure 11.5 illustrates the process. The current capabilities of scanning, artificial intelligence, and rule-base technology are usually not sufficient to completely process the types of documents and records found in most users' environments. Research and

The black-box approach to data conversion.

development of increasingly automated/expert scanning and recognition systems continues. However, most experts believe costs associated with building the expert system rules-base will generally not be competitive in the conversion market unless very large quantities of data are involved.

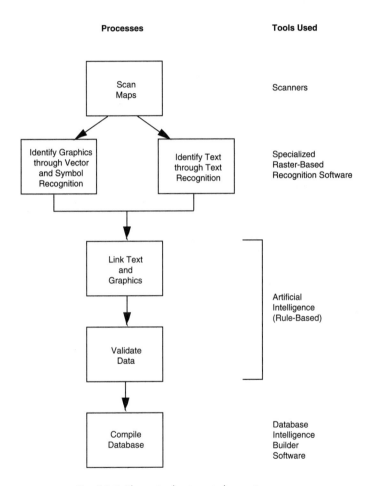

Fig. 11.5. Theoretical automated scanning process.

Interactive Vectorization

Interactive vectorization takes two forms: heads-up digitizing and the interactive use of automated vectorization processes. The advantage of heads-up digitizing is that it is faster for an operator to digitize over a raster line on the screen than to have to continually refocus attention from a paper map on a digitizing tablet to the digitizing display screen to verify data capture. In the interactive process, the raster data is displayed in a background layer on a graphic screen while the operator digitizes vectors on another layer, displayed on the same screen, assisted by automated line-following software. Tests indicate savings in the range of 10 to 20 percent.

Raster Data Utilization

As the data conversion industry closely monitors developments in automated and interactive vectorization technologies, other potential ways to utilize scanning technology in GIS are being investigated. In certain cases, raster data can provide part or all of the graphic information that a GIS database needs. Within an imaging/spatial indexing scheme, raster data may supply an organization with increased production while keeping conversion costs down. However, raster data alone enable a very limited set of applications because they do not provide connectivity or other relational information about the data they contain.

While raster files are large and have substantial storage requirements, technology such as optical storage, file compression techniques, and high-speed workstations is beginning to overcome file size as a limiting factor. Raster data can be utilized in the construction of a GIS in several ways, such as for land base information, reference/detail drawings, and actually as the map itself.

Raster Land Base

Utilizing raster data as a land base is gaining popularity with those users who do not require highly accurate positional data, but only need to visually see the local conditions in the area of their facilities. Most of these implementations take one of two forms. The land base maps on which users wish to locate their facilities can be scanned and tiled to form a completely seamless raster base map. This base map can then be utilized as a background for the display of facilities data generally digitized and stored in vector form on a separate layer.

For users with high positional accuracy requirements, digital orthophotography can be utilized in a similar fashion, with the advantage of having more accurate coordinate locations for each raster pixel. Digital orthophotography provides an alternate and less costly solution to an accurate photogrammetrically compiled vector land base.

Reference/Detail Drawings

Many new generation GIS platforms have integrated the ability to attach raster images of reference or detail drawings to a point on a map. In this way, the significant cost of digitizing those drawings is reduced by offering the ability to view the raster image of the original. For many applications, this use of raster data is cost-effective; the only limiting factors are the ability of the GIS platform to provide an easy to use interface, and the capabilities of the hardware to store, retrieve, and display raster images.

Raster Maps

Utilization of raster images for mapping has been successful as an inexpensive stopgap measure to preserve the information on maps that have deteriorated severely. Archival storage of existing documents by scanning them is a valid alternative to the storage of paper documents.

Incremental Map Conversion

While most conversion processes could generally be considered incremental, since they all approach the data conversion process in a step-by-step fashion, the industry refers to incremental conversion in a particular context. Typically, during incremental conversion projects, all maps are scanned and stored as raster files. Then, as maintenance and construction operations are commenced in a geographic area located on particular maps, the affected maps are vectorized, either by heads-up digitizing or by interactive vectorization. In this fashion, all of the maps will eventually be converted into vector form, leading ultimately to a complete vector database.

Hybrid Files

Recently, this incremental approach has evolved to allow the user to vectorize only a portion of a particular document and to save the file with a portion in vector format and the balance in raster format as shown figure 11.6. While this reduces the vectorization effort associated with the processing of work orders, it may delay realization of the benefits of a completely vectorized database. Additionally, utilization of such hybrid raster/vector files requires additional resources for file management and storage.

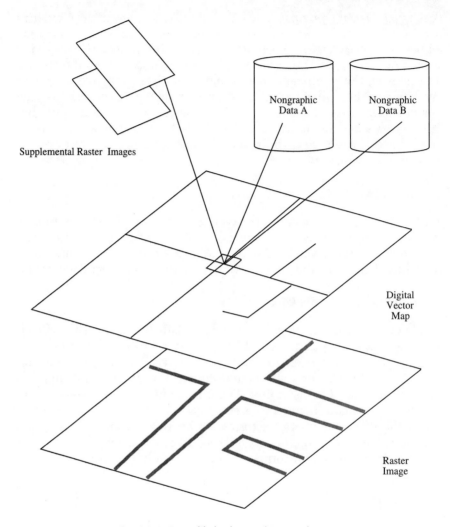

Fig. 11.6. Typical hybrid raster/vector solution.

11.4 Field Data Collection

Many utilities records are so out-of-date that their current users consider them unreliable, and no design work is started without a field visit to determine current conditions. Often the only way to obtain current and accurate facilities data is through field inventory.

Traditional conversion processes that include a field inventory require crews to go into the field with plots of the land base data. The crews will verify or even redraw the entire facilities network onto these plots for entry via traditional tablet digitization. Any existing digital information available from related systems must be incorporated, and discrepancies resolved, before the GIS data can be released to users. The nature of this work requires field crew personnel to have a thorough understanding of the facilities that are being inventoried.

Recent developments in portable and hand-held computers have led to a new approach to field inventory. Many contractors have utilized data collectors with custom software and data entry templates. Conversion vendors have developed proprietary collection software that can be run on inexpensive laptop computers. The evolution of

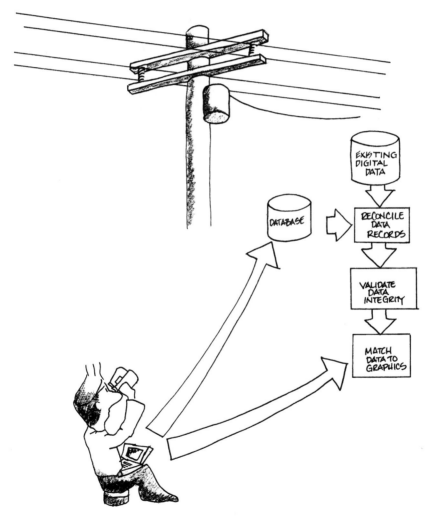

Automated field data collection.

pen-based computing is already improving field data gathering. By using unique feature identification numbers, the inventory data can be recorded digitally in the field and loaded directly into the target system database following a series of range and value checks and a reconciliation process with other digital data.

The advantages of using a computer with database access include increased accuracy, and reduced data compilation effort and costly workstation digitizing time. Mistakes are reduced by eliminating the additional manual steps required to transcribe the data onto a plot and then interpret that information at the digitizing tablet. Off-line edit routines can be run more cost-effectively on the data before loading them into the target system, as can discrepancy checks against existing systems.

11.5 Global Positioning Systems and Conversion

Field data collectors and GPS survey devices have been interfaced successfully. Data collection applications have been developed that utilize the GPS coordinate as the unique feature identification number, enabling a faster and more accurate method of not only recording the facility information but placing it accurately on the land base.

Tools for GIS can be utilized effectively to perform real-time geocoding operations in the field. This alternative has become very valuable, especially since submeter accuracies are now achieved inexpensively.

Additionally, some builders of low-cost land bases are realizing that they can create roadway network maps by utilizing kinematic GPS survey techniques with a roving receiver mounted on a moving vehicle that creates a digital vector file as the vehicle moves down the highway. In the next few years, look for the integration of ever more accurate forms of GPS data into map conversion/creation processes.

In summary, more and more types of computer-based technologies are being investigated to see if they can be used as conversion tools. These emerging technologies are implemented as applicable, either to reduce conversion costs or to compress conversion schedules. Data conversion companies are searching for more effective and less expensive conversion methods, and solution developers are looking for new applications for their technologies, assuring that new potential conversion alternatives are quickly recognized. Therefore, new methods are expected to replace present data conversion procedures fairly quickly.

A good example of this is the use of the open nonsystem-specific databases mentioned in this chapter. The ability to create database intelligence as a precursor to the creation of graphic files during conversion allows important cost reductions and simplifies very complex conversion processes.

Chapter 12

Developing a GIS Data Conversion Plan

12.1 Introduction

The principal asset and characteristic of the most successful GIS data conversion projects is proper planning. Many organizations have rushed into data conversion without proper planning, only to find that they had to throw away valuable resources, including time, and start over. Unfortunately, it is too easy to start a pilot project and use it as a tool that defines GIS project parameters, instead of employing it as a valuable test of carefully planned prototype solutions.

Because of the magnitude and complexity of a typical GIS implementation project, a rushed, unplanned approach can cause problems. The negative results can be grossly delayed schedules, major budget overruns, a database that does not support the anticipated applications, legal action by and/or against conversion contractors, difficulties in finding replacement conversion contractors, rejection of the new GIS by the organization's executive users, and many other technical and organizational problems. There have been several cases where major GIS implementation projects have been cancelled or significantly postponed as a result of poor planning. After an initial failure, an organization's management will be less receptive to (or may even oppose) funding subsequent GIS implementations.

Many different philosophies have been applied to the GIS data conversion challenge, but long-time industry observers tend to agree on one thing: the time and money invested in planning are well spent. After all, data conversion is generally the most costly and time-consuming component of a GIS implementation project. Accordingly, this chapter addresses the subject of getting data conversion under way from a management perspective and examines the most important aspects of GIS data conversion start-up.

12.2 Identifying Sources of Information to Be Converted

The first step in any GIS data conversion project is identifying who will contribute to the database definition on the basis of their knowledge of the organization and its needs. These will be employees who are knowledgeable in the business functions and workflows of the organization and who know where specific source information is located.

These potential users need to be interviewed in detail regarding the data they require to perform their function within the organization, and the data that they desire to support future functions. A compilation of the interview results will serve as the basis for the functional requirements and preliminary conceptual database design. The interview process makes the GIS database design user-driven, and therefore more readily accepted because it will contain familiar data components.

12.3 GIS Database Design Factors

Prior to starting data conversion, several fundamental characteristics of the GIS database must be formulated. Such characteristics include the following:

- Positional accuracy
- Database content and structure (i.e., intelligence)
- Output map product scales

The first consideration in getting started with data conversion involves an assessment of positional accuracy needs. Within a municipal government, for example, this assessment quite often becomes a debate between the engineers, who contend that a GIS database will have little or no value unless it guarantees absolute positional accuracies of ±1 foot or better, and other potential users (e.g., planners) who are satisfied with low positional accuracies on the order of ±100 feet.

To resolve such disputes, the management of an organization, or an outside consultant, should introduce the simple realities of data conversion economics. If nine out of ten departments within an organization are satisfied with a lower positional accuracy and if that lower accuracy costs one-fourth to one-fifth of the high positional accuracy level advocated by the tenth department, it usually becomes incumbent on the tenth department to economically justify and pay the difference. Often, the final positional accuracy level is determined by both democratic and economic principles. Engineering and legal issues may enter into consideration.

Figure 12.1 depicts the relative effect of positional accuracies on GIS data conversion costs. By reducing the allowable positional error from ±100 feet to ±1 foot, the data conversion cost can increase exponentially.

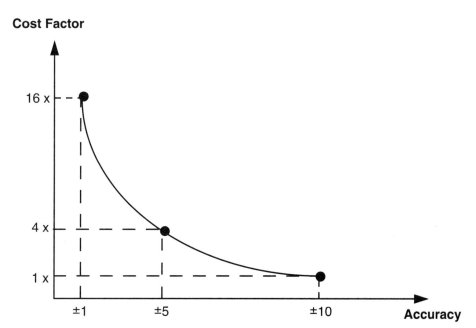

Fig. 12.1. GIS positional accuracy vs. cost.

The other aspect that must be considered in such cases is the positional accuracy of existing source documents. Naturally, wherever economically possible, it is desirable to improve the positional accuracy of maps used by an organization as data are being converted. However, in many cases, cost factors will limit an organization's ability to improve the level of accuracy. When GIS positional accuracy is higher than the positional accuracy of the source documents to be converted, the source documents often must be redrawn onto the new land base using a facility layout process. This layout process is labor-intensive, time-consuming, and expensive. The entire issue of positional accuracy is a balancing act for an organization's management: balancing the overall user needs with the organization's ability to pay.

As discussed in detail in chapter 8, "GIS Database Design," conceptual database design and physical database design are prerequisites to the data conversion process. Organizations must invest the appropriate amount of time and effort to ensure that the database design supports the full range of anticipated GIS applications, or at least that a realistic multistage database evolution is possible.

At the conceptual design stage, it is important that all potential GIS users have an opportunity for input in terms of the features and attributes that they would like included in the database. While generally not feasible to include all suggestions, subsequent exclusion of certain features and/or attributes (perhaps unique to just one department) will spawn less protest since input was requested and objectively considered during the design process. Often features and/or attributes are excluded because of the high cost of data conversion or because they are only occasionally used by the organization.

Data characteristics and the degree of required data intelligence should also be thoroughly evaluated prior to data conversion. The GIS database design being implemented will most likely be in place for 10 to 15 years (or longer) as the backbone of all GIS-related applications for the organization. The approach taken must ensure that the conversion will not have to be repeated.

After review and approval by potential GIS users, the results of the conceptual database design are used for the physical database design, which is the actual building of the GIS database structure on a vendor-specific GIS. A governing principle here (again, from the management perspective) is that the conceptual database design should be carefully examined (audited) before it is transformed into a physical design. This examination ensures that the conceptual database design does not represent more than the organization can afford in terms of initial data conversion, ongoing data maintenance, or both.

Each user department responsible for performing database maintenance for each feature and attribute must be clearly identified. In the harsh economic realities of the 1990s, it is difficult to justify the conversion of an item that cannot or will not be maintained and frequently used in the GIS database. The conceptual database design should, however, contain future features and/or attributes that will not be part of the initial data conversion but will be required to support anticipated future applications.

Prioritizing database needs is often (and predictably) necessary. The database design may include many items that will not be utilized immediately by the organization (e.g., for generating new work orders) and that may not be initially converted by the conversion contractor. Some organizations have implemented multitiered data conversion plans to incrementally populate the database over many years in accordance with a predetermined data item priority schedule.

Output map types and scales are another important database factor. Even if an organization will have on-line access to a GIS, many potential users will still need to obtain a map of the database to use in the field or away from the GIS. Also, most potential GIS users who need maps are accustomed to working with a certain map type and scale where objects are shown at a certain size. Similar to the issue of positional accuracy, there may be a wide divergence of opinion as to the ideal map types and scales to be produced by the conversion contractor as check plots, final plots, and so on.

Again, this is a case of balancing user needs with the economics of data conversion. A recent trend is for organizations to get only the plots needed for their quality assurance tests from the conversion contractor. The remainder of the map types and scales required to support the business functions of the organization are plotted in-house. This process helps minimize conversion contractor plotting charges.

Another example of map plotting costs is illustrated by the case where one department requests 1"=100' (1:1,200) check plots and another requests 1"=200' (1:2,400) plots. Certainly, one of the basic advantages of GIS over manual drafting methods is the ability to produce maps at various user-defined scales. However, in this example, the total plotting costs for the 1"=100' (1:1,200) plots will be significantly higher than for the smaller scale maps. This is because a 1"=100' (1:1,200) map covers only one-fourth the geographic area that a 1"=200' (1:2,400) map covers.

12.4 Other Factors

From a management perspective, the goal of a GIS is to realize an overall productivity improvement in conjunction with other tangible benefits. Especially during data conversion start-up, technical improvements that do not fit within the GIS project scope should be considered secondary objectives to the implementation project.

Political factors can, and do, influence the prioritization process. This can be very pronounced on GIS projects that are authorized by an approving body (e.g., city councils or executive committees). To make the best possible impression on approving bodies during initial GIS demonstrations, a project manager may give highest data conversion priority to the items most likely to "win over" borderline dissenters.

12.5 Identifying Existing Data Sources

It is imperative that the parties responsible for preparing the initial data conversion cost estimates and other planning activities conduct a thorough analysis of the existing maps and documents that will be utilized as input to the data conversion process. Any manual card files, tabular records, existing digital data, or other available information that could be of value in the data conversion effort must also be included in this preconversion analysis. The fundamental objective of the analysis is to prevent surprises later.

The analysis of potential data conversion sources includes matching the needs of the database design with the availability of source data. For example, if a particular feature included in the physical database design has nine associated attributes to be converted by the conversion contractor, a data source(s) must be identified for each of the nine attributes.

Moreover, it must be confirmed that the source identified for each attribute covers the organization's entire area of interest. Many times, a single data source is identified as the sole source of information from which to populate a certain database record field, but later it is discovered that the source is only available for a portion of the area of interest. A scramble then ensues to determine a suitable replacement source. Obviously, data conversion schedules can be jeopardized while such unanticipated analyses are conducted.

12.6 Identifying New Data Sources

Another key aspect of getting started in GIS data conversion is the identification of source data that has to be created. As part of this effort, it is important to consider the introduction of new potential sources for items beyond the obvious ones that may have been originally envisioned. For example, an electric utility may elect to have new aerial photography flown to serve as the source for the majority of land base features. However, these photographs (if flown at appropriate flight height) may also be used to locate poles, padmounted transformers, and some other utility features. The photographs may be a better source for the conversion of these items than existing distribution maps or pole cards.

Once again, prioritizing such data items and preferred source decisions is related to economics. Stated another way, every available piece of existing geographic information is rarely, if ever, converted into the GIS database. Some data sources examined during the analysis phase will be determined to be obsolete. Others will be determined to be so incomplete or unreliable as to not warrant conversion. The objective is to identify the key, reliable, and current or new data sources that can be used to construct the foundation of the GIS.

The range of GIS data to be converted varies dramatically among GIS projects. Some organizations require only a moderate level of relative positional accuracy. Such organizations may elect to acquire existing government data as the basis for certain land information. Other organizations require a high degree of absolute positional accuracy, and new ground surveys may be required to support the subsequent aerial mapping operations. In these cases, GPS techniques are often employed (refer to chapter 6, "Data Conversion Methods") to acquire new positional reference data. A GIS data conversion project really has two components: acquisition and conversion, as seen in the example of GPS and aerial mapping.

Similarly, on the facilities side, a data acquisition component may exist. Analogous to a field survey or GPS survey to acquire new (improved) land information, a field inventory may be utilized to gather and record information in support of the facilities data conversion effort. Field inventories are used as a tool to field check source documents before data conversion and to collect missing or new information.

The need for such land and facilities data acquisition activities as prerequisites to the data conversion of existing source records must be evaluated based on each GIS project's individual technical and management objectives. However, even in cases where a substantial amount of data acquisition is necessary, the costs for this activity rarely exceed 40 percent of the total data conversion costs. Ten percent to 15 percent of the total data conversion cost is typical.

12.7 Conceptual Database Design

From a management point of view, the manner in which the conceptual database design effort is conducted is a subject of discussion in the industry. Some advocate a conceptual database design approach that not only includes the precise inventory of data features and attributes to be contained within the GIS database but also addresses and thoroughly documents the required database relationships.

To a large extent, this debate can be settled by knowing when a specific GIS will be selected. In most GIS experts' view, it is a waste of time to attempt to design and document very detailed data relationships in a situation where the conceptual database design work is being conducted prior to the selection of a vendor-specific GIS. Certainly, some of the design decisions made concerning specific data relationships will depend on the capabilities of a specific GIS. Conversely, if the system of choice is confidently known prior to commencing conceptual database design activities, then, by all means, the design should include detailed database relationships. Doing so will expedite and simplify the downstream physical database design task.

Also, an organization must be able to handle specific applications, and these applications will have an impact on the type and brand of

GIS that will be selected. Therefore, the conceptual database design must include unique data relationships that are necessary to support the specific applications to be implemented by an organization. In any case, the conceptual database design process should be done in line with the overview presented in chapter 8, "GIS Database Design."

From a careful examination of user requirements, a framework of anticipated GIS applications can be developed. This framework drives the identification of features and attributes to be included in the database, and lays the foundation for the development of application software.

Often, populating the GIS database (even in temporal form) is best accomplished by using existing digital data or by implementing real-time interfaces to an organization's other computer systems. The conceptual database design must include the data structure necessary to support interfaces to other systems.

The conceptual database design is a precursor to the physical database design process. Physical database design generally culminates with the coding of a project-specific database schema that reflects the myriad design decisions made along the way.

Conceptual database design generally requires two to four months, depending on the number of user reviews, the impact and timing of GIS selection, the number of other tasks going on at the same time, and a number of other variables. Physical database design can take anywhere from one week to many months, depending on the database design complexity, the ability of the GIS vendor to utilize a previous similar design, and several other factors.

The organization's GIS users must constantly review the progress of the conceptual database design process to make sure that they have made the desired modifications, additions, and deletions prior to the physical database design effort. Their review should also include decisions in terms of source materials, costs, records preparation requirements, and so on.

12.8 Physical Database Design

In the 1970s and early 1980s, the GIS vendor had a significant role in physical database design. This was viewed as necessary since GIS vendors were familiar with the most effective ways to model facilities networks or land-related information topology under their proprietary data structures, with the system architecture restrictions affecting database design on their hardware platform, and with performance considerations relating to applications and specific design decisions. End users were not normally qualified to perform the task of vendor-specific physical GIS database design. In the 1990s, the role of the GIS vendor in physical database design will diminish due to the advent of RDBMS and the presence of experienced RDBMS database designers within organizations implementing GIS. Vendor input will still be solicited, but the actual work will most often be done by the organization.

Conversion contractors can also contribute to the physical database design process, especially when they have production experience with the vendor-specific GIS used by the organization. The ideal combination of personnel to be involved in the physical database design effort includes experienced technical representatives from the GIS vendor, the conversion contractor, representatives from the organization implementing the GIS, and a GIS consultant. Due to the practical limitations of contract timing, however, this ideal situation rarely occurs. The conversion contractor is often absent from the initial physical database design phase.

As discussed in chapter 8, "GIS Database Design," a GIS user review of a preliminary version of the physical database design is an important subtask in the data conversion start-up effort. This review is usually the user's first opportunity to see the end result of the user requirements surveys and the conceptual database design.

The media used to present the physical database design for review varies. Some organizations prefer to have a paper copy of the physical database design to facilitate group review meetings and to serve as tangible GIS project documentation. Others are satisfied with the digital database schema which is displayed and reviewed at a GIS workstation.

Coordination of the physical database design is crucial to other activities. Most conversion contractors are unwilling to commence any data conversion tasks beyond ground control acquisition and the flying of aerial photography without the benefit of a final physical database design.

Other activities, such as initial hardware installation, are more independent of the physical database design step. Lengthy periods between major steps on the path to GIS project implementation should be avoided. For example, it would not be wise to complete end-user training months before the physical database design was ready for review because the training will have been forgotten long before the GIS can be used productively.

12.9 Internal vs. External Conversion

Thus far, the discussion related to getting started with a GIS data conversion project has, for the most part, assumed that organizations will engage external contractors for the bulk of data conversion and that the organizations will not perform much of the initial data conversion internally (themselves). This is due to the fact that most organizations, after carefully considering the resources required, timing, expertise, and cost factors, generally elect not to get into the data conversion business just to accomplish their individual GIS project implementation. Most organizations that are starting a GIS implementation do not have the in-depth knowledge, the surplus personnel, the equipment, or the custom software to do data conversion efficiently and/or correctly. In addition, conversion contractors are usually forced to work very cost-effectively within contractual schedule requirements. Project schedules and budgets are usually hard to achieve internally in an organization that is just starting a GIS learning curve.

Certainly, there are exceptions. Several organizations have elected to conduct their data conversion in-house. These organizations found the longer conversion schedules often associated with internal efforts acceptable. They also determined acceptable ways to manage a potential overabundance of both equipment and manpower upon completion of the data conversion effort.

12.10 Purchasing Data Conversion Services

The procurement of GIS data conversion services has evolved into a relatively refined and systematic process. The entire process is a highly competitive one. Sole source data conversion contracts were somewhat common in the 1970s and 1980s but have been rare in the 1990s.

To begin the process, a request for information (RFI) is typically issued by an organization to several conversion contractors who are presumed to have the interest and necessary qualifications. An RFI is sometimes referred to as a request for qualifications (RFQ), and the response from the contractors is referred to as a statement of qualifications (SOQ). The trend has been toward the term RFI and away from the term RFQ, because the request for qualifications acronym gets confused with the accepted business use of RFQ—request for quotation. This is an important distinction for GIS data conversion, since an RFI is not intended to request cost information.

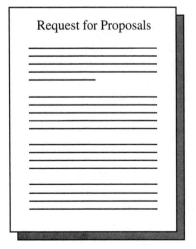

After careful evaluation of the RFI responses, a shortlist of conversion contractors is usually developed. The two to four firms on this list are those that have the necessary background, expertise, resources, and demonstrated interest to do the work at hand. Accordingly, these firms receive a subsequent request for proposals (RFP).

Developing proposals is a very involved task since it normally involves a comprehensive set of data conversion specifications. It is not unusual for the overall development process to require three to five months. However, as figure 12.2 illustrates, this task is usually accomplished at the same time as several other data conversion project start-up tasks.

It is common for the organization soliciting the data conversion proposals to host a prebid conference with potential conversion contractors. This conference is generally scheduled on a date one to two weeks following the release of the RFP. Recipients of RFP are invited to submit written questions concerning the GIS data conversion project prior to the prebid conference.

During the prebid conference, the organization will typically provide additional information concerning the GIS implementation project in general, provide answers to the previously submitted written questions, conduct an open question and answer session, afford the attendees the opportunity to inspect additional data conversion sources, etc. In some cases, the conversion contractors will request a separate (more lengthy) period of time for source inspection/analysis. This is generally agreed to; however, some organizations refuse because of the additional time demands placed upon their staff, or because of procurement policies that prohibit such visits after issuance of the RFP.

Proposals from the conversion contractors are generally due four to eight weeks after an RFP is released. The evaluation of proposals for data conversion services is a complex process. The most effective way to conduct the selection process is to utilize a mathematical scoring model during the evaluation. In terms of the time investment, it is common for an individual scorer of conversion proposals to spend one to two days evaluating each conversion contractor's technical proposal and another one-half day to evaluate each price proposal. Price proposals should remain sealed until after the technical scoring has been completed. This requirement will help to ensure that the technical scores are not biased by knowledge of the contractors' prices.

A thorough evaluation of conversion proposals will also include a detailed look at risk factors relating to each conversion contractor's stability, size, experience, project management track record, financial health, litigation history, and so on. The risk assessment portion of the evaluation process also entails diligent checking of each conversion contractor's client references.

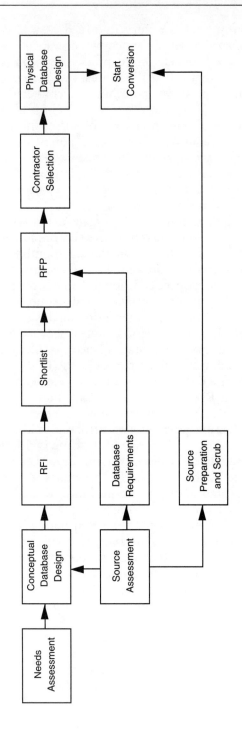

Fig. 12.2. GIS database purchase cycle.

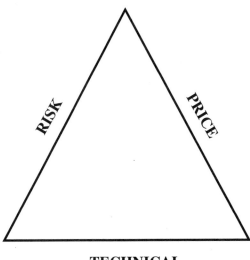

Thus, the selection of a competent conversion contractor involves a decision triad composed of technical, price, and risk scores. The relative weighting of each of these triad scores varies among organizations. Consultants generally recommend a formula that places emphasis on the technical evaluation. However, a variety of sensitivity analyses utilizing different weighting are often performed prior to the final selection of the preferred conversion contractor.

The process of signing a contract for GIS data conversion services depends, to a large extent, on the purchasing philosophy employed in the development of the original RFP package. In some cases, the organization will include a standard contract in the RFP that reflects their standard terms and conditions. In other cases, the RFP solicits a sample conversion contract from the contractors.

Under either scenario, so-called standard contracts that are applied to engineering, construction, or other routine business activities seldom satisfy the unique conditions, issues, and requirements of GIS data conversion. An organization preparing to place a conversion contractor under contract should plan on investing the effort necessary to tailor the contract wording to topics directly applicable to data conversion. The conversion RFP and the conversion contractor's proposal should be addenda to the contract.

The initial contract generally applies only to the data conversion work for a pilot area, with the understanding that the prices and procedures that are confirmed will then extend to the full area of interest (assuming a positive outcome from the pilot). Through careful RFP and contract development, most organizations can avoid the need to release subsequent RFPs for data conversion services following the pilot. However, some organizations actually prefer to periodically release RFPs, each covering the next increment of data conversion work.

Opponents of this approach argue that the organization potentially ends up paying for multiple conversion contractor project start-up costs. On the other hand, organizations that advocate multiple RFPs argue that they do not want to be locked into, or totally dependent upon, a single conversion contractor for the entire data conversion project. The strength of these respective arguments is often a function of the total project size and project schedule. Multiple conversion contractors may be necessary to complete the data conversion in a short time frame.

12.11 Source Preparation and Scrub

Source document preparation and scrub procedures seldom receive the attention and resources they deserve during the data conversion start-up period. If the conversion contractor personnel cannot read the information on the client's maps and records, the converted GIS database will be correspondingly poor and/or data conversion costs will escalate.

The specific sources requiring scrub need to be identified early. The responsibility for scrub operations needs to be clearly defined because most scrub operations can be handled by either the conversion contractor or the client.

Often, a formal manual that addresses the scrub plan and detailed scrub requirements is a recommended starting point. This manual is usually produced by the same personnel involved in the initial conceptual database design and data conversion specifications development.

Information that needs to be added, corrected, or reformatted during data conversion is identified by source document and in the terminology of the physical database design (i.e., correct feature and attribute names). In any case, the definition of the scrub procedures should be driven by database and application requirements.

This manual must be readily available and understandable to all members of the GIS project team—not just the project manager. The project manager is responsible for ensuring that the manual includes appropriate measurement criteria and a scrub budget.

Since determining scrub responsibilities is a vital step in the process, the level of scrub effort on each source document must be estimated, and a systematic approach must be developed for each of the client's departments involved in the scrub process. Timing and coordination with other tasks are also considerations here, as shown in figure 12.3. The plan must be detailed enough to include specific personnel assignments, anticipated scrub time for each source type (for client-conducted scrub activities), and an overall scrub schedule.

12.12 GIS Data Conversion Work Plan

Another critical component of data conversion start-up is the development of a comprehensive work plan. This plan concentrates on the production side of data conversion. It addresses the following topics:

- Client GIS user personnel
- Conversion contractor personnel
- Lines of communication
- Source document preparation
- Source document reproduction
- Scrub timing relative to previous steps
- Size of source packets (quantity of facets, etc.)
- Size of deliverable packets (quantity of check plots, area covered, etc.)
- Frequency of deliveries
- Project schedule
- Project budget
- Summary of reporting requirements

The GIS user personnel portion of the data conversion work plan should contain an organization chart and accompanying descriptive narratives. At a minimum, the chart illustrates and the narratives describe the roles, functions, and reporting relationships among the following personnel for the data conversion project:

- Project manager
- Database administrator
- Source preparation and reproduction staff
- Data acceptance staff (quality assurance personnel)

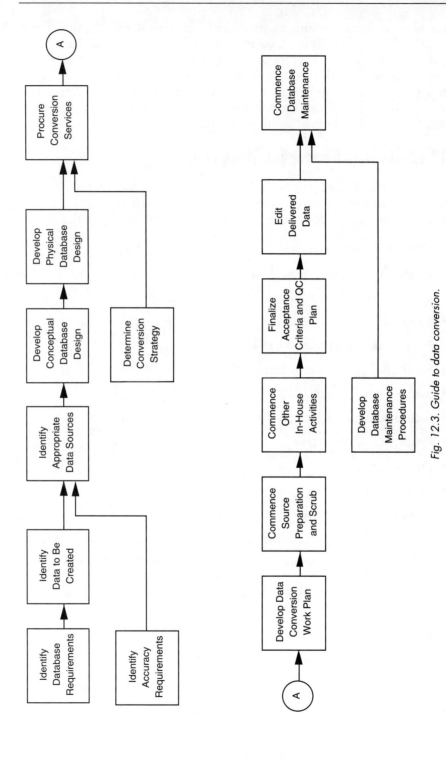

Fig. 12.3. Guide to data conversion.

- Technical development staff (analysts and programmers)
- Accounts payable

For most GIS projects, it is also appropriate to include within this section of the work plan the interface points between the client's personnel and their counterparts at the conversion contractor. This would, of course, be expanded to include the quality assurance contractor in a scenario that involves a third party to conduct quality assurance and data acceptance activities. The data conversion work plan contains a chart and accompanying description for individuals involved at the conversion contractor site(s). If subcontractors are involved in the effort, they are included here as well.

The data conversion work plan also includes a clear description of the lines of communication between the client and the contractor(s). Although the original RFP indicated a need for frequent client/contractor meetings, the details surrounding such meetings are itemized within this section of the work plan. A preliminary plan showing meeting locations, frequency, expected attendees, and expected duration should be formulated at this time. Almost certainly, changes will occur as the pilot progresses, and a strong contact between the parties will greatly reduce misunderstandings and problems.

Other forms of communication, such as facsimile transmissions of source document questions, status reports, billing questions, and so forth, should also be documented in the data conversion work plan. However, work plan emphasis remains on the lines of communication. The document should spell out who, from each party involved, deals with the various types of issues expected to arise during the data conversion process.

On very large GIS data conversion projects, there are commonly three levels of communication. The client's project manager or database administrator deals with the conversion contractor's project manager on issues concerning acceptance of deliverables, overall project schedule, and other general project concerns. The client's technical staff deals with their respective counterparts at the conversion contractor on database design, hardware, and software issues. The conversion contractor's designated production representative (sometimes known as a conversion coordinator) interacts with the client's personnel responsible for conversion specifications, source document preparation, and other topics. However, any specification changes must be controlled by the project managers.

Full agreement and understanding among the parties on the lines of communication must be established. Without such common understanding, the data conversion project can become bogged down with inappropriate problem resolutions caused by improper communications. Also, database design decisions that may seem to enhance an application may have an adverse impact on data conversion if not cleared at the proper level before being implemented.

Source document reproduction should be addressed in detail within the data conversion work plan. Each type of source information involved in the data conversion effort must be included. The data conversion work plan clearly defines the media required for the source copies, the number of copies of each source required (usually only one, but can be multiple), the required contrast/legibility of each source type, the physical size of the copies, and any other information related to the reproduction process. More fundamentally, the responsibility for reproducing the sources must be clearly defined. While it is most common for the client to make the copies (as indicated in the conversion RFP), there are cases where the conversion contractor is responsible for supplying the personnel, equipment, and material necessary to produce all required source copies.

Scrub methodology and scheduling is often covered in a subsection separate from source document preparation. For example, the reproduction of source documents not requiring preparation can occur at any time prior to shipment to the conversion contractor. Other source types should not be copied until the preparation/scrub operations have been completed. This part of the data conversion work plan identifies the sequence of preparation/scrub/reproduction for each source document type. The extent of scrub to be done internally and externally is typically not fully defined until the conversion contractor is selected. A certain level of backlog must be maintained to avoid idle periods for the various conversion processes.

The data conversion work plan also specifies the size or quantity of the group of sources to be delivered to the conversion contractor as a work packet. Usually, the conversion source documents are delivered in manageable numbers corresponding to a predetermined geographic area. For example, a utility covering thousands of square miles will not send all of the source documents for their entire service territory in a single large shipment. Instead, they will prepare a conversion work plan that assumes a certain conversion rate (in square miles per week). The assumed conversion rate is used to determine the corresponding source work packet size and delivery rate at the front end of the data conversion process.

On the other end of the process, the size of converted work packets (data tapes and check plots) is also specified within the data conversion work plan. From the client's perspective, the size of the incoming converted packets may or may not have a one-to-one correlation with the size of the source work packets. This varies from project to project and is a function of staffing at each end and the conversion contractor's production rate. The key is for the client to determine a converted packet size that does not overtax the company's ability to inspect data for acceptance, but does meet the specified delivery schedule.

The combination of the size of the converted packets and the overall data conversion schedule will help establish the frequency of the deliverables from the conversion contractor. Biweekly or monthly deliveries are fairly common, regardless of the client's size. Weekly deliveries tend to generate excessive overhead costs in clerical time, preparation time, and in express shipping charges.

A detailed project schedule is included in the data conversion work plan. Because of the political pressures on many GIS data conversion projects, this can often be the most important section in the plan.

The schedule should lay out each source work packet delivery to the contractor and each converted packet shipment to the client in terms of start and end dates for each group of packets. Packets are generally grouped logically by department, division, district, area, etc. The schedule is most effectively created and maintained using a computerized program capable of producing horizontal bar schedules. In addition to production-related tasks, other important milestones are depicted on the schedule (e.g., training, hardware installation, or budget cycles).

For a pilot data conversion project, a sampling of project schedule task titles (durations would be shown for each on the actual schedule) includes the following:

- Contract signing.
- Each ground control task.
- Each aerial photography task
- Photoprocessing
- Analytical triangulation
- Physical database design
- Conversion software customization
- Stereocompilation (may separate planimetric versus topographic)
- Client collection and preparation of source documents
- Copy sources
- Contractor's preparation tasks
- Each digitizing operation (may be source dependent)

- Check plot production
- Translation software development
- Pilot area delivery
- Client review
- Client pilot acceptance
- Contractor corrections
- Client specification changes
- Physical database refinements and conversion contractor software changes

A condensed version of the project budget is generally included in the data conversion work plan. By condensed, it is meant that project cost categories such as initial feasibility studies, hardware procurement, software procurement, consultants' implementation support, and other nonconversion-oriented items are excluded. Projected monthly expenditures are itemized by project deliverables according to the schedule.

Some organizations choose to include other costs such as personnel training, recruitment, and relocation (for those individuals whose focus is data conversion). Others record and track these costs under budgets separate from the data conversion budget.

In any case, it is important to ensure that the proper GIS data conversion budget foundation and cost tracking mechanisms are in place to evaluate progress against the original budget and to detect cost overruns and other budget-related problems early enough to take effective corrective action. From the project manager's point of view, it is generally advisable to have the data conversion budget include not only the dollars flowing to the conversion contractor but also the internal costs (or at least the time) associated with source preparation, source reproduction, data loading, quality assurance and acceptance, and any other client costs related directly to data conversion.

As referred to in the discussion on lines of communication, the data conversion work plan should document the project reporting requirements. This is usually accomplished at a summary level. Status reporting requirements are listed and a sample report is developed and included in the plan.

Status reports from the conversion contractor are generally a contract topic and should include status information concerning work in progress, work completed, and future tasks. The conversion contractor's status report should include a list of problems, resolutions, and open issues. The format of such reports is mutually agreed upon, and breaks down subdivisions into discrete tasks and subtasks. Early or late task completions are also included. A section for issues and a separate section for comments are integrated into the form design. The data conversion work plan also specifies the frequency of reporting (e.g., a weekly facsimile transmission).

12.13 In-House Activities

Beyond what may be contained in the formal work plan described above, the client's in-house activities need to be identified, agreed upon, documented, and implemented. The coordination of all in-house activities is no small undertaking. Failure to adequately provide such coordination is one of the quickest ways to have a data conversion project acquire a negative reputation within the organization.

Similar to the project management and reporting tools designed to track the progress and problems of the conversion contractor, a set of effective tools needs to be implemented for in-house project management. These tools are often designed to address and control the many data management issues that begin to surface as soon as the first converted packet delivery is received. For example: Is the land base data delivered and approved prior to the facilities data? How are new deliveries merged with previous deliveries? How often are magnetic tape backups run? What tracking mechanisms are used to distinguish accepted data from rejected data? When are errors corrected, and when are they returned to the conversion contractor? How is the turnaround time for rejected data tracked? Who has access to which data? How are updates handled? While some conversion purists argue that many of these concerns are outside the realm of data conversion, the fact remains that a failure to adequately address these issues (and many others) can adversely affect the entire data conversion process.

12.14 Quality Assurance

As mentioned in the preceding section, there is an obvious need for the conversion client to accurately determine the acceptability of the delivered data. The process and procedures to accomplish this task must be worked out and documented during the start-up period. Acceptance criteria details can be very complex. Proper analysis and resultant documentation is meant to ward off issues surrounding the counting of errors, what constitutes rejection, consistency between client tallies and conversion contractor tallies, and a common understanding of rejection and correction tracking.

For the acceptance checking process itself, random sampling techniques are often used to avoid conducting a detailed check of each delivered data set. An option that is rapidly gaining popularity is for a neutral third party to conduct the quality assurance operations. During project start-up, full 100 percent quality assurance is typically done to verify accuracy and to find problem areas. Sampling may later be implemented, with emphasis on critical GIS database items.

Generally, the client (or their designated quality control agent-contractor) will have a specified period (usually 30 to 60 days) following the receipt of the converted packet delivery to accept or reject each discrete deliverable. The conversion contractor will have a lesser amount of time, usually 15 days, to correct and redeliver any rejected products.

One of the most interesting aspects of the quality assurance process is the interpretation and flexibility of contract terms related to data acceptance criteria. Rarely does an end user enforce their data rejection rights 100 percent. For example, there are times when a facet of data is discovered to be outside acceptable tolerance by a slim amount. In some of these instances, the end user may determine that it requires less time and effort to correct the product in-house than to incur the administrative time in rejecting and subsequently rechecking the product after the contractor's corrections are made. This is a very subjective call and is, therefore, almost never formally documented. Project managers usually give their quality assurance personnel some general guidelines under which to operate.

Payment for accepted data is usually made within 30 days following acceptance and is occasionally subject to a withholding (also known as holdback or retainage). Such retainage generally does not exceed 15 percent, with 10 percent being considered typical. The release of the retainage varies. Some contracts call for the release of the retainage at the completion of the contract; others specify release upon acceptance of the next converted packet; others tie it to the achievement of an overall data conversion project completion percentage.

In the event that the organization has elected to utilize a neutral third party for quality assurance and acceptance checking purposes, the data conversion start-up period will include the related procurement and negotiation activities. In some instances, this includes contracting with a surveying firm to perform positional accuracy checks on the ground for specified database objects.

12.15 Change Control

Although an ideal data conversion project requires absolutely no changes once the database design and data conversion specification molds are cast, changes are to be expected in many aspects of the GIS data conversion endeavor. Therefore, part of a comprehensive start-up plan includes proper mechanisms to identify, convey, approve (or reject), and implement potential changes to the following:

- Technical specifications
- Database design
- Conversion procedures
- General scope of work/budget

Generally, a form will be devised upon which suggested changes can be identified. The completed form for a particular requested change will preferably include a discussion of the ramifications. Quite often, a requested change must not only be evaluated from a forward-looking perspective (i.e., data yet to be converted) but also take into account the impact on data that has already been converted. The conversion contractor will also record estimated cost/savings associated with the change.

The requested change is usually evaluated by the client who may also enlist the assistance of the GIS vendor and/or a consultant to determine the ultimate disposition of the request. If approved in writing by the client and conversion contractor, the change will be implemented by the appropriate party(s).

Most GIS data conversion RFPs and contracts now acknowledge the inevitability of change requests. For each new data conversion project, however, the task remains to define and enforce the use of the project-specific change control mechanisms.

12.16 GIS Data Maintenance

Naturally, once the converted data is delivered to and accepted by the client, the process of maintaining the GIS database begins. To prepare for maintenance activities, a clear delineation of specific responsibilities must be established and communicated to all concerned.

In the past, some organizations embarking upon a GIS implementation have made the mistake of delaying database maintenance activities. In a few extreme cases, the magnetic tapes from the conversion contractor have hung untouched in a data vault for two or more years.

This is usually a very costly mistake. First, the entire acceptance process can be jeopardized if the converted data is not immediately loaded and tested upon receipt. Second, the problem of outdated information, which was most likely a key motivation for implementing GIS in the first place, can quickly resurface.

Detailed GIS data maintenance procedures are also required. In some cases, GIS consultants are utilized to assist in this effort. Vendors and conversion contractors also have been known to assist end users in developing data maintenance procedures.

Conversion contractors also regularly perform GIS database maintenance for clients until the client's personnel are trained. Under this scenario, the contract for database maintenance is usually separate from the original conversion contract.

In terms of preparing a GIS database maintenance schedule, the organization should plan for database maintenance to commence on the first day after a converted packet is accepted. To accomplish this, hardware and software installation, personnel training, and other related activities must be coordinated with the quality assurance process. The plans must also take into account the fact that maintenance will increase as the amount of accepted converted packets increases.

In addition to the procedures previously mentioned, other forms of database maintenance documentation are generally developed. Forms (either paper or on-line) are developed to track information such as operator's name, type of updates made, job or project numbers, etc.

Other forms of temporary documentation may be required to assist the organization during the transition period from the manual system to the GIS. Most departments will be forced to live with a dual system during the data conversion period. That is, they will be performing manual maintenance activities on the yet-to-be-converted sources as well as computerized maintenance via the GIS on the data that have been converted and accepted. This requirement for parallel database maintenance activities can cause significant problems if not carefully planned for and managed.

12.17 Pilot Project

The concept of a pilot project has been repeatedly mentioned throughout this publication. When viewed as a GIS project implementation phase, it is arguably the single most important step in the overall data conversion project. The pilot project typically embodies all of the post-database design efforts that have been the subject of this chapter.

The basic concept of a pilot area is to convert a small geographic area to enable the client, the GIS vendor, and the conversion contractor to meet certain objectives. Typical pilot objectives to be accomplished include the following:

- Test database content
- Test suitability of sources
- Test database structure
- Test document preparation and scrub activities
- Test data conversion procedures
- Confirm project-specific symbology
- Test quality assurance procedures
- Test data acceptance procedures
- Test pilot applications
- Confirm data conversion cost estimates/budget
- Provide a first major milestone for the GIS implementation process

The development of a pilot plan generally happens concurrently with the development of the data conversion specifications. The pilot plan amplifies each of the mentioned objectives and clearly delineates responsibilities. A line style chart (often produced with common project management software) provides an excellent way to combine all aspects of the pilot project plan onto a single chart and schedule.

A properly developed pilot plan will also satisfy another need. It will take the users of the intended GIS through a very valuable learning curve, allowing them to take the next step: a full implementation of the GIS. Apart from learning how to proceed in technical terms, the user will be able to apply his or her newly acquired analysis and planning tools to approach the implementation of an overall GIS solution much more effectively and successfully.

Glossary

Word(s) or phrases in **boldface** are defined in this glossary.

Absolute orientation
Scaling, leveling, and orientation to ground control of a stereo pair of aerial photographs during the photogrammetric setup process.

Acceptance criteria
In data conversion contracts, the measures of data quality used to determine whether conversion work has been performed according to specifications.

Acceptance tests
(1) A test that evaluates a newly purchased system's performance and conformity to specifications. (2) A test performed by a user or user organization to check that a delivered system component (hardware, software, or data) meets the agreed upon requirements.

Access time
A measure of the time interval between the instant that data are called from storage and the instant that delivery is complete.

Accuracy
Degree of conformity with a standard or accepted value. Relates to the quality of a result and is distinguished from precision, which relates to the quality of the operation by which the result is obtained.

Accuracy, absolute
The degree of difference in the position of a map feature from its true position on the ground, based on accepted coordinates established through geodetic survey of control points.

Accuracy, relative
The degree of difference in the scaled distance between two map features from the corresponding ground distance between the actual features. Differs from **absolute accuracy** because the accuracy of the position of a given feature is based on the difference between its map distance from another feature and the corresponding ground distance; absolute accuracy is based on a map feature's distance from its own true location on the ground.

Acetate overlays
A nonflammable plastic sheeting used as a base for photographic films or as a drafting base for overlays where critical registration is not required.

Address
(1) A number referring to a location in computer memory. (2) A street location.

Address matching
The ability to match an address component to its geographic location on the ground.

Admatch
The process by which geographic records containing street address information are assigned additional information about geographic areas that pertain to that address, usually for the purpose of aggregating information by geographic area, for example, admatching to determine census tracts and blocks or voting districts for individual households. See also **address matching**.

Aerial photographs
Photograph of the earth's surface taken with an aircraft-mounted camera.

Aerial triangulations
A method of establishing coordinates for points that are apparent on an aerial photograph, based on calculation of distances and angles from other apparent points whose coordinates are known. Also referred to as aerotriangulation.

Affine transformations
Coordinate transformation to convert map coordinates into database coordinates.

Algorithms
A set of rules or procedures for solving a specific mathematical problem.

Alias
An alternate name for a graphic entity. Often used to record various names for streets and roads (e.g., "Interstate 25" and "Valley Highway").

Alphanumeric database
A collection of data consisting of alphabetical characters and numeric information. Distinct from a graphics- or spatial-oriented database.

AM/FM
See **automated mapping/facilities management.**

AM/FM International
A nonprofit, educational association that fosters information exchange, educational opportunities, and scientific research and development to advance and promote geographic and facilities management information systems.

American Society for Photogrammetry and Remote Sensing (ASPRS)
A professional society dedicated to promoting/sharing photogrammetry and remote sensing technology.

Analog
Representation of a numerical quantity by continuously variable physical qualities such as graphic marks or electric voltages; contrasted with digital.

Analog stereoplotter
Mechanical-optical device used to create maps from aerial photography.

Analytical stereoplotter
Software controlled device used to create highly accurate three-dimensional models from aerial photography.

Analytical triangulation
Process of densifying the ground control network so that enough ground control points are available for setting up each stereoscopic model during photogrammetry.

Annotation
Text on a drawing or map associated with graphics entities.

Apparent centerline
The centerline of a feature, such as a street, as perceived from aerial photographs.

Applications
Routines utilizing geographic information to perform some type of analysis or report; typically used in the decision, designing, and planning processes. Application software programs generally refer to an addition to the inherent capabilities of the base system software. These programs are either developed by internal systems personnel or purchased from the software vendor or a third party.

Applications development
Programs either developed by internal systems personnel or purchased from the software vendor or a third party to supplement application software program capabilities.

As-built drawings
Engineering drawing that shows the placement of facilities as measured after work is completed. In some cases, significantly different from the design drawings for a facility.

Aspect
The horizontal direction in which a slope faces.

ASPRS
See **American Society for Photogrammetry and Remote Sensing**.

Attributes
Descriptive information about each land or facilities element in the database. Examples would be material, height, treatment, inspection date, ownership, etc., for a utility pole.

Automated conversion
The process of converting maps and drawings into a digital format while reducing or eliminating operator intervention. Scanning is one form of automated conversion. Heads-up digitizing via tracing a raster image displayed on a computer screen is another form.

Automated mapping
The use of computer graphics technology to produce maps. May include specialized **CAD** technology for digitizing and editing map features, and **DBMS** technology to produce thematic maps.

Automated mapping/facilities management (AM/FM)
Specific segment of the computer graphics industry concerned with specialized geographic and facilities information management and applications. Especially appropriate for public and private utilities, city and county governments, and other organizations that depend on accurate, up-to-date geographic and facilities information. Computer graphics technology provides the capacity to enter, store, and update graphic and nongraphic information. The primary goals of most AM/FM systems are to contain the cost of maintaining and using geographic and facilities information, to create and store geographic and facilities information in a digital database, and to make quality geographic and facilities data available to users.

Axis
A reference line in a coordinate system.

Azimuth
(1) Horizontal direction of a line measured clockwise from a reference place, usually the meridian. (2) A way to measure directions in degrees of an angle. Measured most of the time from the north; the military, however, measures it from the south. Always measured in a clockwise direction and never exceeds 360°.

Backlighting
During digitizing, refers to using digitizing tables which illuminate the map from below.

Base maps
A map containing visible surface features and boundaries, accurately referenced to a specific coordinate system.

Baseline
(1) A starting point from which future improvements will be compared. (2) Set of costs projected from the base year over the study period.

Batch process
A collection of programs stored in a queue for later processing (noninteractive).

Benchmark
A monument (i.e., marked) point on the ground whose elevation above a reference surface (e.g., mean sea level) is known.

Benchmark tests
Various standard tests, easily duplicated, measuring product performance under typical conditions of use.

Bit
Abbreviation for **bi**nary digi**t**; a number that can take only the values zero or one.

Bit maps
A pattern of bits on a grid stored in memory and used to generate an image on a raster display.

Blocks
A group of words or records treated as a logical unit of information.

Boolean operations
Allows selection of items in a database using set theory concepts of union and intersection, in terms of expressions which include AND, OR, and NOT. For example, select all properties with street address on Main Street AND zoned residential.

Breakline
Break in the terrain that is either natural (such as ridges and drains) or man-made (such as road edges and retaining walls).

Buffer
An internal, temporary memory that provides intermediate storage between the **central processing unit** and the disk or printer.

Buffer zone
Generation of a polygon around a feature, where sides are defined by a given distance from the feature. Also referred to as corridor generation.

Bulk loading
Process of entering data in a database from a preexisting data file.

Bytes
A group of bits that can be stored and retrieved as a unit.

Cable throw
Telephone terminology for the function of moving a group of working telephone circuits from one set of cable pairs to another in the same cable or to another cable. Typically done to provide service to clients.

CAD
See **computer-aided drafting.**

Cadastral maps
A graphic representation of a portion of the earth's surface that shows the delineation of parcels of land and indicates the relative size and position of each parcel in relation to other properties, roads, streams, and other major physical and cultural features. Drawn to an appropriate scale and displays dimensions and areas with identifying parcel numbers.

Cadastral property overlays
Recorded acceptable plots of land parcels, when available, plotted on a true copy of a base map with proper adjustments to reconcile any discrepancies.

Cadastral survey
A survey that creates, marks, defines, retraces, or re-establishes the boundaries and subdivisions of the public lands of the United States.

Cadastre
(1) A survey that creates, defines, retraces, or reestablishes the boundaries and subdivisions of public lands and private estates. Ownership and values of private lands are recorded for taxation. (2) Pertains to legal definition of property boundaries, ownership, etc.

CADD
See **computer-aided design and drafting.**

CAE
See **computer-aided engineering.**

Cartesian coordinates
A coordinate system in which the locations of points in space are expressed by reference to three planes, called the coordinate planes (X,Y,Z), no two of which are parallel.

Cartographic accuracy
The degree to which a map measurement is known to approximate a given value.

Cartography
The science and art of making maps.

Cathode-ray tube (CRT)
An electronic tube with a screen, which is used with computer terminals to display letters, numbers, symbols, and graphics.

CCITT Group IV
International data compression standard commonly used in the production of communications equipment such as facsimile machines.

CD-ROM
See **compact-disk, read-only memory.**

Cells
The basic element of spatial information in the raster (grid) description of spatial entities.

Census blocks
Usually a small area (city blocks) used to compile the population count for a larger given area.

Census tracts
An area made up of several census blocks for compiling periodic governmental enumerations of population.

Central processing unit (CPU)
The portion of the computer that controls the hardware (screen, printer, disks, etc.) and completes tasks assigned by a program.

GIS Data Conversion Handbook

Centroids
(1) A point in the interior of a polygon, often used to identify the polygon. (2) The point whose coordinates are the average values of all the coordinates that define the figure.

Characters
A letter, symbol, or digit; usually a byte.

Choropleth maps
A map that shows statistics related to geographical areas by shading the areas based on statistical values.

Client load
Peak cumulative demand on an electrical utility system at any given time.

COGO
See **coordinate geometry.**

Colinearity equations
Set of equations used by analytical stereoplotters to create three-dimensional stereomodels.

Compact-disk, read-only memory (CD-ROM) drive
A storage device, similar to an audio compact disk player, capable of reading disks that hold approximately 500 megabytes of digital data.

Compilation scale
Scale at which a map is created.

Completeness
Measure of the degree to which all features are included in a database as a result of conversion.

Computer-aided design and drafting (CADD)
Involves the computerization of design processes in addition to CAD. Recently, the term *CAD* has begun to mean both CADD and CAD.

Computer-aided drafting (CAD)
An application of computer graphics technology that automates manual drafting techniques.

Computer-aided engineering (CAE)
The integration of computer graphics with engineering techniques to facilitate and optimize the analysis, design, construction, operation, and maintenance of physical systems.

Connectivity
The ability to trace from a source to any given point. An example is an electric distribution system where each individual circuit can be traced.

Continuous map
A digital representation of a land area that is unbroken by map borders or boundaries, as opposed to separate, individual maps.

Contour
A line that connects points of equal elevation.

Contour interval
Elevation change between two consecutive contour lines.

Contouring
Stereoplotter technique where the measuring mark is set to a constant elevation value and then moved along the ground at that elevation in the stereomodel.

Control network
Lines connected to form a system of loops or circuits extending over an area (not boundary or map sheet oriented).

Control points
A point whose coordinates are known. Used as a reference for other surveys.

Conversion
The process of transforming existing manual information from its present form (maps, drawings, records, etc.) to a digital format, typically in a database stored on a computer.

Conversion services firm
An organization having AM/FM/GIS equipment that assists in or provides conversion services, commonly referred to as a service bureau or conversion vendor.

Conversion specifications
Requirements that are used to guide the conversion of data.

Coordinate geometry (COGO)
A data entry method that accepts data in the form of survey data (i.e., bearings and distances) and calculates coordinates for points.

Coordinate system
A particular kind of reference frame or system, such as plane rectangular coordinates or spherical coordinates, that uses linear or angular quantities to designate the position of points within that system.

Coordinate transformation
Mathematical calculation that transforms data represented in one coordinate system or projection to another (e.g., latitude-longitude to UTM coordinates, UTM coordinates to state plane coordinates). Often a component of a GIS.

Coordinates
The positions of points in space with respect to X, Y, and Z axes. *See* **Cartesian coordinates**.

Corridor generation
See **buffer**.

Creators
Individuals within an organization who are responsible for creating document sets for their own use or for the use of others.

Credibility
In cartography refers to the degree to which map users trust and believe the information contained on the map.

CRT
See **cathode-ray tube**.

Cursor
A visible symbol guided by a keyboard, joystick, tracking ball, or digitizer, usually in the form of a cross or a blinking symbol, that indicates a position on a CRT.

Dangle
When a graphic line/element does not connect properly from digitizing. The line is digitized past its intersection with another line.

Data
(1) A general term used to denote facts, numbers, letters, and symbols that refer to or describe an object, idea, condition, situation, or other factors. (2) Connotes basic elements of information that can be processed or produced by a computer.

Data elements
Individual items, features, or components in a GIS database. Examples are streets, poles, transformers, zoning districts, etc.

Data encoding
Converting data to machine-readable format.

Data entry
Process of recording data into a computer.

Data field
The smallest unit of discrete data that can be accessed within a data record.

Data file
See **data set**.

Data format
The way in which data elements are represented and stored in computer records.

Data layers
Grouping of geographic data that emulates, in concept, the organization of information on map overlays.

Data migration
The process of moving data from one system to another, often from a system being decommissioned to a new system.

Data quality
Composite of data completeness, data correctness, and data integrity.

Data records
A set comprised of associated data fields. For example, a group of related statistics, a fixed or variable length machine-readable record.

Data segmentations
Data organization, usually into layers or levels, but increasingly into objects.

Data set (data file)
A collection of stored data records.

Data sources
Map or document from which data will be collected for inclusion into an AM/FM/GIS database.

Data translation
The process of converting data from one form to another.

Data vintage
A measure of when data were collected or verified.

Database design, conceptual
The process of defining data requirements, data organization, and data relationships, without regard to a specific hardware platform or software environment.

Database design, physical
Actual definition of the database schema using a particular **DBMS** product.

Database management systems (DBMS)
An organized set of data that allows flexible query and manipulation of tabular data by selecting portions of the entire data set for further calculation, analysis, and reporting.

Database structures
Information about data organization and data relationships. Also referred to as schema.

Databases
A collection of interrelated data sets stored together and controlled by a specific schema. A consistent and specified set of procedures is used in adding data to the set and in changing or retrieving existing data.

Databases, hierarchical
A database where data are linked together in a tree-like fashion, similar to the concept of a family tree, where relationships can be traced through a particular pathway of the hierarchy. As opposed to relational data, hierarchical data files are not independent and the data structure must be known in order to extract information.

Databases, relational
A database scheme by which information stored on a computer can be input and retrieved independently from other related data. Has its strength in its ability to relate information easily and quickly and, most important, transparently to the user. Disadvantages are performance and retrieval speed. The major advantage of the relational database scheme is its flexibility and versatility by virtue of the fact that many different types of data are stored independently of one another, yet can be easily retrieved and linked without knowledge of the data structure.

Datum
Any numerical or geometrical quantity or set of such quantities which may serve as a reference or base value for other quantities.

DEM
See **digital elevation model**.

Densification
Process of adding control points to a control network in order to increase the attainable accuracy of the depending databases.

Destination systems
The system on which data is intended to be used. Often differs from the system used in the data conversion process.

Diapositives
A photograph positive on a transparent medium, usually polyester or glass.

Differential leveling
A method of vertical surveying involving the use of a leveling instrument and graduated rods.

Digital data
Data that exists in a format that can be used and accepted by a computer. In GIS, digital data are generally divided into two categories: graphic and attribute (nongraphic).

Digital elevation model (DEM)
A file with terrain elevations recorded for the intersection of a fine-grained grid and organized by quadrangle as the digital equivalent of the elevation data on a topographic base map.

Digital holography
Technology which, when fully developed, will provide three-dimensional models of actual field conditions for design, planning, and troubleshooting purposes.

Digital image processing
Technology that utilizes remotely sensed geographic information represented as an image in digital raster format, and allows classification of various land covers and ground features based on characteristics of the data.

Digital line graph (DLG)
Digital representations of **USGS** topographic quadrangle maps. The DLG format is a standardized file format that retains spatial relationships between all map features (**points**, **lines**, and **polygons**), and that contains attributes of features such as highway classification and type of water body. Digital line graph data is available for **USGS** map products at 1:250,000, 1:100,000, and 1:24,000 (excludes topographic, i.e., elevation information).

Digital orthophotographs
Commonly referred to as orthophotos; produced by scanning aerial photographs and using elevation data to perform image rectification on the scanned photographs. The resulting digital data can be printed to produce traditional orthophotos or viewed directly on a display screen.

Digital terrain models (DTM)
A three-dimensional view of a section of the earth's surface.

Digitize
One means of converting manually prepared maps and geographic records into digital data that can be accepted and used by a computer. Used to denote the act of drawing or tracing with a computer, using a **cursor**, **digitizing tablet**, and **CRT**.

Digitize, manually
The process of converting an analog map or other graphic overlay into numeric format with the use of a digitizing table/tablet and manually tracing the input data with a cursor.

Digitizers
Equipment used to perform digitization. Also used to refer to a person who performs digitizing work.

Digitizing tables
Large format digitizer (D-size, E-size, and larger). Typically mounted on a special stand.

Digitizing tablets
Small format digitizer (usually up to C-size). Designed to be placed on the work table near the keyboard.

DIME
See **dual-independent map encoding**.

Dimensioning
A function of **CAD** systems that computes and inserts the dimensions of a design on the **CRT** screen. The design can then be output as an engineering drawing.

Displacement
The shift in position of a feature on a photograph due to tilt, scale change, or relief.

Displayable attribute
Descriptive information that can be displayed on a screen or positioned on a map adjacent to the element it is associated with.

Displays
To present the information stored on a system in visible form.

DLG
See **digital line graph**.

Document imaging
The technology of electronically capturing, storing, distributing, annotating, displaying, and printing documentation previously available only on paper.

Document set
A consistent group or series of maps, drawings, or records that have the same subject matter, format, and purpose.

Dots per inch (dpi)
Common measure of printer/plotter resolution. Generally, the higher the number of dots per inch (dpi) the better the quality of the output.

Downloading
Process of extracting data from a host computer.

dpi
See **dots per inch**.

Drawings
For an AM/FM/GIS, graphic depiction of land and facilities; may or may not be related to specific geography.

Drawing layers
Groups of similar data related to drawings. Layers are useful for quickly identifying groups of features to be used in a particular drawing or analysis and also for protecting groups of features from access by other users.

DTM
See **digital terrain model**.

Dual-independent map encoding (DIME)
U.S. Census Bureau computer files created for urbanized areas in the United States for the 1980 Census, containing digital street map information. The DIME file structure explicitly retains spatial relationships between street segments and the blocks they form. Commonly used as a reference file for **address matching** (ADMATCH) operations. Was superseded by **TIGER** in 1990. Also known as geographic base file (GBF)/DIME.

Dynamic segmentation
A model for storing attributes of linear features where the attributes do not correspond to the same portions of the linear feature. The actual position of the attribute or segmentation can change dynamically without impacting other segmentation in the data model.

Eastings
Distance along the X axis in rectangular (**Cartesian**) coordinate systems used for geographic positioning.

Edge matching
The process of identifying a common point where a line or symbol crosses a shared border of two adjacent maps or drawings.

EDM
See **electronic document management**.

Electronic document management (EDM)
The technology of electronically capturing, storing, distributing, annotating, displaying, and printing documentation previously available only on paper.

Electrostatic edit plot
Plot produced on an electrostatic plotter and used for quality control.

Elements
Individual items, features, or components in an AM/FM/GIS database. Examples are streets, poles, transformers, switches, etc.

Ellipsoid
A surface whose cross-sections are all ellipses or circles, or the solid enclosed by these surfaces.

Encoder
Equipment that converts data into machine-readable format.

Entity
A graphic representation of geographic features. *See* **features**.

Erasable disk
Storage device similar to a **WORM** drive but which allows data to be erased.

External conversion
Data conversion performed using an outside conversion firm.

Facilities
Major asset elements, that when connected together, usually make up the network of an AM/FM/GIS. Examples are an electric operating system, a gas delivery system, and a street/highway system.

Facility data
The major physical elements that usually make up the network data of an AM/FM/GIS when connected together. Examples include poles, cables, sewer pipes, transformers, terminals, and other assets.

Facility models
The combination of facilities that depict a particular network with associated engineering data and attributes that are typically used for network analysis. Usually associated with providing client service or rectifying network analysis issues.

Features
The basic geographic element in a GIS used to represent physical and cultural features on the earth. All planimetric geographic features can be represented in two dimensions as **points**, **lines**, or **polygons**.

Feature attributes
Also called a *feature object*. An element used to represent the nonpositional aspects of an entity.

Federal Geodetic Control Committee (FGCC)
Government organization that determines standards for geodetic surveying.

FGCC
See **Federal Geodetic Control Committee**.

Fiducial marks
Marks which are exposed on film during aerial photography and which provide a reference for photo coordinates to be used during aerial triangulation and orientation in a stereoplotter.

Field inventory
The manual effort of compiling/verifying facility assets or other land oriented facilities located in the outside world (field).

Files
An organized collection of **records**.

Film flattening
Feature of aerial photographic cameras which ensures that the film is flat against the focal plane of the camera.

First-order work
The designation given survey work of the highest prescribed order of precision and accuracy. Formerly called *primary*.

Flatbed plotters
A digital plotter upon which the output material is mounted on a flat surface.

Flatbed scanners
Scanner in which the document is placed on a flat glass surface while the scanner's sensing device moves across the image.

FMC
See **forward motion compensation**.

Foot
One English foot equals 0.3048 meters.

Formats
The arrangement of data within a field, record, or file. Also the type of geographic representation; for example, polygon for lines or grid for grid cells.

Forward motion compensation (FMC)
Ability of aerial photography cameras to slightly advance the film while it is being exposed, thus compensating for the forward motion of the airplane.

Geo-location
Geographic location.

Geocoding
The activity of defining the position of geographical objects relative to a standard reference grid.

Geodetic controls
See **control points**.

Geographic coordinate system
The spherical coordinate system of latitude and longitude that allows unique identification of any location on earth.

Geographic data
A collection of data which are individually or collectively attached to a geographic location. **Spatial data** is a term used synonymously with *geographic data*. Higher nominal location identifiers or specific location identifiers are used to indicate positions in space relative to other data. Nominal location identifiers are typically names or code numbers of geographic entities such as administrative districts, postal zones, street addresses, political subdivisions, rivers, highways, census tracts, and so forth, and arbitrary identification numbers assigned to individual graphic entities such as **points**, **line** segments, and **polygons**. Specific location identifiers are related either to some coordinate system or to other data elements by one or more spatial languages. Four types of location identifiers are used:
Points (as abstractions of small phenomena or surrogates for larger phenomena)
Line segments (for linear features)
Arbitrary (user-defined) *regular areas* (used for processing) or *data management convenience* (for grid cells, pixels)
Irregular polygons that describe surface conditions (for soils, vegetation)

Geographic information systems (GIS) technology
System of computer hardware, software, and procedures designed to support the capture, management, manipulation, analysis, modeling, and display of spatially referenced data for solving complex planning and management problems.

Geographic records
A collection of data that is individually or collectively attached to a geographic location, including the attributes of the entity and its recorded geographic location.

Gigabyte
Approximately 1 billion bytes.

GIS
See **geographic information systems technology**.

Global positioning system (GPS)
A three-dimensional surveying system based on radio signals from the **NAVSTAR** constellation of earth orbiting satellites. Implemented by the U.S. Department of Defense and available for civilian applications on a selective basis.

Graphic displays
The presentation of data on a **CRT** screen, such as a map, drawing, or chart in graphic form so that the data can be interpreted visually.

Graphic editing
The manipulation of graphic data, including repositioning, deleting, copying, rotating, etc.

Graphic elements
Arcs, lines, points, curves, symbols, circles, or other graphic figures and alphanumeric characters displayed on a graphic display or plotted on various media.

Graphic records
Digital information used to represent a feature in a data set or separately as a symbol or template to which individual data records may refer.

Graphics tablets
A surface through which coordinate points can be transmitted by identification with a cursor or stylus. Used synonymously with digitizing table.

Grids
(1) A network of uniformly spaced horizontal and perpendicular lines which enclose an area (cell) with an associated value assigned. (2) A network of uniformly spaced points or lines on the **CRT** for locating positions.

Ground control
Points on the ground used to verify vertical and horizontal accuracy.

Heads-up digitizing
Process of digitizing a map by tracing selected, appropriate portions of a raster image displayed on a **CRT**.

Horizontal control datum
A geodetic reference for horizontal control surveys. Five quantities are known: latitude, longitude, azimuth of a line from the point, and two constants that are the parameters for the reference ellipsoid.

Horizontal control points
A survey station whose position has been accurately determined in x- and y-grid coordinates, or latitude and longitude.

Hydrography
The mapping of water courses and bodies of water.

Hydrological
The class of GIS data features representing water courses and bodies of water.

Hypsography
That part of topography dealing with the relief or elevation of the terrain. Reference is made to a horizontal datum, usually mean sea level.

IGES
See **initial graphics exchange standards**.

Image interpretation (correlation)
The automatic (system generated) matching of position and physical characteristics between imagery of the same geographic area from different types of points of view. This is used by a few very advanced photogrammetric devices for automated aerial mapping.

Image processing
Reproducing representations of objects electronically or by optical means on film, electronic display devices, optical disks, or other media.

Impact printers
Printers that produce characters by hitting the paper with a mechanical printer head to create single characters.

Incremental conversion
Conversion that is phased in over a selected period of time with the intent of incrementally building a database. Criteria must be established for a conversion program to maximize payback. Conversion criteria may be based on geographic areas, specific database features, or perceived applications.
Phrase also applies to conversion programs featuring scanning of source materials followed by heads-up digitizing of features based on selection criteria as described above.

Initial graphics exchange standards (IGES)
Defines neutral data between computer graphics systems.

Integrity
A measure of data quality that focuses on the relationships among data elements.

Intelligent infrastructure
The process of using modern computer graphics technology integrated with advanced database management systems to plan and manage spatially linked facilities and land records systems. Other technologies comprise intelligent infrastructure. These technologies include **automated mapping/facilities management** systems (AM/FM), **geographic information systems** (GIS), and **land-related information systems**. In addition, intelligent infrastructure systems manage work processes that deal with infrastructure information.

Interactive
Refers to a system allowing two-way electronic communication between the user and the computer.

Interactive graphics
Capability to perform pictorial operations directly and interactively on the computer screen with immediate feedback and results.

Interior orientation
Factors such as focal length, lens distortion, principal point locations, etc., which determine the precision of aerial cameras.

Internal conversion
Data conversion performed by in-house staff and/or by using in-house equipment.

Interpretive photogrammetry
The use of aerial photographs to identify objects and determine their significance.

Key entry
Using an alphanumeric keyboard to enter commands and attribute data into a system. Key entry is also used to enter precise coordinates, distances, bearings, and azimuths during **COGO** functions.

Land base
Group of maps covering a specific geographic area on which facility information is overlaid. Can be either paper-based or digital.

Land-related information systems (LRIS)
Computer-based system used to manage geographically-linked, nonfacilities, land-based information.

Landsat
A series of NASA remote sensing satellites designed to acquire data about the earth's resources.

Large scale
A map scale that covers a relatively small area on the ground and has features shown in detailed labels. The term large refers to the fraction represented by the ratio of map distance to ground distance. For example, 1:500 (one map unit=500 ground units).

Layers
Logical concept used to distinguish subdivided group(s) of data within a given computerized map or drawing. May be thought of as a series of transparencies (overlaid) in any order, containing information pertaining to a specific set of related data such as an electric distribution system. A workstation operator may specify display elements (layer) to be visible (on) or invisible (off).

Leveling
Measuring vertical distances, directly or indirectly, to determine elevations.

Light pens
A hand-held photosensitive interactive device for identifying elements displayed on a light sensitive **CRT** screen.

Lines
(1) One of the basic graphic elements, defined by at least two pairs of X, Y coordinates. (2) A level of spatial measurement referring to a one-dimensional defined object having a length, direction, and connecting at least two points. Examples are roads, railroads, utility lines, streams, etc.

Line snapping
Software function that forces a line to end exactly at another **linear features**.

Linear features
Features that can be adequately represented with **lines**.

Linen
High-quality paper used in map production.

Locations
Position of an object or surface feature on the face of the earth expressed in terms of coordinate points.

Lots
A measured parcel of land having fixed boundaries and designated on a plot or survey.

LRIS
See **land-related information systems**.

Macros
A text file containing a series of frequently used operations that can be executed by a single command. Can also refer to simple high-level programming languages with which the user can manipulate the commands in a **GIS**.

Magnetic disks
Storage device that uses electromagnetic mechanisms to record data.

Mainframes
Large multiuser computer usually optimized for transaction processing and management of extremely large databases.

Management information systems (MIS)
A computer-based system designed to provide the management of information necessary for day-to-day operations in an organization. System personnel comprise a group of individuals responsible for an organization's central data processing functions and services.

Manholes
An underground structure providing access to utility facilities.

Manually digitize
See **digitize, manually**.

Maps
Graphic representation of the physical features of a part of or the whole of the earth's surface by means of symbols or photographic imagery at an established scale, on a specified projection, with orientations indicated.

Map features
The representation of an object on a map.

Map projections
An orderly system of lines on a two-dimensional plane, representing a corresponding system of imaginary lines on an adopted surface, such as latitude/longitude lines on the earth's surface.

Map scales
The relationship existing between a distance on a map and the corresponding distance on the earth.

Megabyte
Approximately one million bytes.

Menus
A list of options on a display allowing an operator to select the next operations by indicating one or more choices with a pointing device.

Menu-driven
Operated from a menu.

Metric photogrammetry
The making of precise measurements from photographs to determine the specific location of objects.

Minicomputers
Midsize multiuser computer.

Models
The mathematical representation of a process or system that can be manipulated to show the effects of certain actions on the process or system.

Monocomparators
Instrument used to measure photo coordinates from one photograph at a time.

Monumentation
The establishment of a permanent marking of public land survey corners and fixing line positions so that the location of the surveyed lands can be precisely known.

Multiparticipant GIS projects
A **GIS** project involving several agencies.

Mylar™
A brand of strong, thin, and transparent polyester film used in photography and map production.

NAD 83
See **North American Datum of 1983**.

Nadir
The point on the ground directly beneath the perspective center of a camera lens, in aerial photography.

National map accuracy standards (NMAS)
(1) Horizontal accuracy: For maps at publication scales larger than 1:20,000, 90 percent of all well-defined features will be located within 1/30 inch of their geographic positions. (2) Vertical accuracy: Ninety percent of all contours and elevations interpolated from contours will be accurate within one-half of the contour interval.

NAVSTAR
The constellation of satellites that comprise the navigation and **GPS** with which a three-dimensional geodetic position and the velocity of a user at a point on or near the earth can be determined in real time.

Neat lines
The line on the edge of a map where graphic detail stops.

Networks
The characteristic of an **AM/FM** or **GIS** whereby spatial relationships, or connectivity of points along a linear feature are retained within the database. This allows the system to automatically trace a path between an origin point and a destination point. Applications that typically require this type of spatial relationship include routing applications, sewer and water network analyses, and so on.

Network pair connectivity
A telephone term that refers to the connectivity of the numerical sequences of cable pairs within a cable network. Typically from a central office to a subscriber for the purpose of providing service.

Network topology
Data relationships built to track connectivity in a network.

NMAS
See **national map accuracy standards**.

Nodes
A common point between two or more line segments.

Nongraphic tabular data
Alphanumeric data that is stored or can easily be displayed in tabular form.

Normalize
To allow comparisons between incompatible data.

North American Datum of 1983 (NAD 83)
The horizontal control datum for the United States, Canada, Mexico, and Central America, based on a geocentric origin and the Geodetic Reference System 1980. The basis for all maps created since 1986 (a small percentage of the total).

Northings
Distance along the Y axis in a rectangular coordinate system (**Cartesian**).

OCR
See **optical character recognition**.

Off-line
The transmission of information between a computer and a peripheral unit before or after, but not during, processing, in contrast to **on-line** processing.

Off-line text entry
Process of entering data into a file which will later be posted to the database.

Offset line
A supplementary graphic line close to and parallel with a main line, to which it is referred by measured offsets.

On-line
The transmission of information between a computer and a terminal or display device while processing is occurring, in contrast to **off-line** processing.

Optical character recognition (OCR)
Technology that allows scanned text to be recognized as individual characters.

Orientation
The act of establishing correct relationship in direction with reference to the points on a compass.

Orthophotographs
A photograph prepared from aerial photography that has been adjusted to show objects in their true geographic position by removing displacement caused by camera tilt and topographic relief.

OSP
See **outside plant**.

Output scale
The scale at which a hard copy of a map is produced.

Outside plant (OSP)
A general term applied to all physical facility property of a utility, typically a telephone company, that contributes to the furnishing of client service.

Overlaps
That portion of a map or photograph that overlaps the area covered by another of the same series; common area covered by adjoining aerial photographs.

Overlays
Data layer, usually dealing with only one aspect of related information, that is used to supplement a database. Overlays are registered to the base data by a common coordinate system.

Oversheets
A transparency or a print of a map used for recording supplemental information.

Overshoot
That portion of a graphic line that extends past another intersecting line. *See* **dangle**.

Pan
Graphic display functions that allow the user of a **GIS** to get selective views of different areas of a display by moving left, right, up and down, or across an image.

Paneling
The placing of markers on the ground and the marking of surface features such that they will be visible on aerial photographs. Also called *targeting*.

Parallax
Apparent displacement of an object caused by its position of observation.

Parcels
The fundamental unit of land; the basic building block for maintaining land information, including the information about rights and interests.

Parcel mapping
The mapping of land parcels.

PC
See **personal computers**.

Personal computers (PCs)
General term for microcomputers. Often refers to IBM or IBM-compatible microcomputers.

Photogrammetry
A process of determining the location of photo-identifiable points on the surface of the earth by the use of **aerial photographs** and **ground control**.

Photograph control
See **paneling**.

Photograph interpretation
The act of examining photography images for the purpose of identifying objects and judging their significance.

Pilot projects
A limited application of a GIS project used for testing preliminary design assumptions, data conversion strategies, and system applications. A pilot project is usually conducted for a small portion of the project's geographic area.

Pixels
(1) Picture element. The smallest discrete element that makes up an image. (2) Smallest unit of information in a grid cell map or scanner image.

Planimetric map
A map that shows the horizontal (x,y) positions of features (i.e., no topographic features). Planimetric features commonly include streets, buildings, rivers, lakes, etc.

Planimetry
The natural and man-made features represented on a map with the exception of relief.

Plotting
Producing a hard copy of a map on a pen or raster plotter.

Points
A level of spatial definition referring to an object that has no dimension. Map examples include wells, weather stations, and navigational lights.

Point-in-polygon operations
Polygon overlay operation where point features that lie within polygon features are counted, displayed, or otherwise selected or analyzed. An example of this analysis is counting the number of fire hydrants within a fire district.

Pole cards
A record used by utilities to keep track of their pole facilities in the field with associated information such as height, class, treatment, value, and year placed.

Polygon intersection
The result of two or more polygons overlaying forming a new common area. *See also* **polygon processing**.

Polygon overlay
See **polygon processing**.

Polygon processing
A general term for analytical operations of geographic information systems involving the overlaying of various separate layers of geographic information to form new information. Polygon operations include point-in-polygon analysis, line-in-polygon analysis, and polygon overlay operations of dissolve and merge. The results of these operations create new layers of information based on the combined characteristics of the input layers. These operations require the system to retain information on spatial relationships between all map features.

Polygon retrieval
The ability to extract data contained within irregular boundary lines (tax zones, rate areas, meter reading areas, etc.) on a map.

Polygons
A two-dimensional figure with three or more straight-line sides (arcs) intersecting at a like number of points or nodes. Area polygons may refer to parcels, boundaries, serving districts, etc.

Positional accuracy
Overall reliability of the positions of cartographic features relative to their true position, or with respect to each other.

PPS
See **precise positioning services**.

Precise calculations
The maximum possible refinement of reference points by utilizing all intelligence sources and analytical computer techniques.

Precise positioning services (PPS)
The most accurate dynamic positioning possible with **GPS**, based on the dual frequency P-code.

Precision
A statistical measure of repeatability that is usually expressed as variance or standard deviation (root mean square or RMS) of repeated measurements.

Precision placement
The placement of geographically referenced object or information using a statistical measure of repeatability, usually expressed as a variant or standard deviation of repeated measurements.

Premarking
Process of painting targets on ground features such as manholes and street intersection monuments to make them easier to see on aerial photographs.

Primary sources
Document used as the initial and principal source for data conversion.

Profiles
A vertical section of the surface of the ground, or of underlying strata, or both, along any fixed line.

Publication scales
Output scales.

Pucks
The hand-held input device used for digitizing.

Pugs
Tiny hole drilled in the emulsion of diapositives for use during **aerial triangulation**.

Quadrangles
Four-sided area, bounded by parallels of latitude and meridians of longitude, used as an area unit in mapping. Usually refers to USGS topographic maps.

Qualitative benefits
Benefits that improve the quality of a service or a product but that are not easily measured.

Quality controls
An aggregate of activities designed to ensure adequate quality, especially in the conversion of manual records into GIS data.

Quantitative benefits
Tangible or measurable benefits.

Radial displacements
The shift in position of the image of an object on a photograph measured along a line drawn from the center of the photograph through the image of the object.

Raster data
Cell data arranged in a regular grid pattern in which each unit (or cell) in the grid is assigned an identifying value based on its characteristics.

Raster GIS
A **geographic information system** in which map layers are represented in mosaics of grid cells. This can give the appearance of a somewhat abstract map; however, it lends itself to efficient analysis of data and direct utilization of remotely sensed data. These types of systems have been widely used in natural resource GIS applications. Also known as *grid-cell GIS*.

Raster-to-vector
The process of converting an image made of cells into one described by **lines** and **polygons**.

Records
A collection of associated fields in a file that are treated as a unit. For example, all data items for a census tract can be grouped as a record and assigned to a single segment of a magnetic tape file for convenient storage and retrieval.

Records posting
Updating records in a database to reflect changes made since the last update.

Reflective scanner
Scanner in which the light source and the sensor are on the same side of the document.

Registration
The aligning of map sheets and overlays based on common coordinate points.

Relational databases
A method of structuring data in the form of sets of records or sets of elements so that relations between different entities and attributes can be used for data access and transformation.

Relative accuracy
Measure of the maximum deviation of an object represented on a map from other objects represented on the same map.

Relative orientation
Positioning of a stereo pair of photographs in relation to each other.

Relative positioning
A technique by which two sets of GPS receivers simultaneously measure a known geodetic point and a point to be determined.

Remotely sensed data
Digital land data that is collected via scanners located on stations remote from the earth, usually on satellites. Common remotely sensed data sets include **SPOT** and **Landsat**.

Request for information (RFI)
Requests qualifications information from vendors of AM/FM/GIS services and equipment. The request can take several forms. First, a formal, written document which describes project objectives and requirements, similar to an **RFP**. Second, a formal letter, containing concepts and general guidelines can be prepared. Personal contact, though not generally considered an RFI, achieves the objective of gathering qualifications information.

Request for proposals (RFP)
A formal document issued by an organization requesting interested vendors to submit proposals responding to a predefined set of requirements. Vendor responses take the form of an itemized description of the technical approach to be used, a performance schedule, and a cost schedule with supporting qualifications, capabilities, etc.

Resolution
Measure of the sharpness of an image. Usually expressed in **dots per inch**.

RFI
See **request for information**.

RFP
See **request for proposals**.

Rotating drum scanners
Type of scanner in which the document is mounted on a rotating drum with the light source located at the center of the drum.

Rubbersheeting
A procedure to adjust graphic elements in a from-and-to manner within an area, forcing elements to fit.

Rule-based systems
Systems that operate on the basis of predefined rules of behavior.

Satellite imagery
Image data collected by satellites.

Satellite Probatoire pour l'Observation de la Terre (SPOT)
The French earth observation satellite system featuring a linear array sensor, push-broom scanning, and pointable optics. Full-scene stereoscopic imaging from parallel satellite tracks is available.

Scales
The ratio of a distance on a photograph or map to its corresponding distance on the ground. Scale may be expressed as a unitless ratio (1:10,000), a unitless fraction (1/10,000), or an equivalent with units (1 inch=833.33 feet).

Scan lines
The narrow strip on the ground recorded by the field of view of a scanner system (e.g., a remote sensing satellite).

Scanning
(1) The process of taking hard copy and making an image of the document for use in a computer system. (2) The process of using an electronic input device to convert analog information from maps, photographs, or overlays into a digital format usable by a computer.

Schematic drawing
Drawing that shows functional rather than spatial relationships among objects, or a drawing that has no spatial reference.

Scribe coat
An opaque plastic coating (on a transparent base) into which lines and other symbols can be cut. The finished scribe sheet can be used as a photographic negative.

Scrub
The manual preparation and edit verification/correction of a map or drawing prior to digitization (e.g., incorrect street name, or wrong location number for pole).

Seamless database
A continuous database.

Section corner
The corner at the extremity of a section boundary (subdivision corner).

Side lap
Overlap between adjoining strips of aerial photographs.

SIF
See **standard interchange format**.

Slope maps
A map depicting topographic elevations and information used for analysis (i.e., to calculate the slope gradient of each cell in percent slope).

Small scale
A mapping scale that covers a relatively large area on the ground and has generalized labels.

SOQ
See **statement of qualifications**.

Spatial analysis
Analytical techniques associated with the study of the location of geographical entities together with their spatial dimensions. Also referred to as *quantitative analysis*.

Spatial data
Data pertaining to the location of geographical entities together with their spatial dimensions. Spatial data are classified as points, lines, areas, or surfaces.

Spatial topology
Relationships such as adjacency and connectivity among spatial elements.

SPOT
See **Satellite Probatoire pour l'Observation de la Terre**.

Spot elevation
A point on a map whose elevation is noted.

Spot height
Elevations data taken at specific points.

Standard interchange format (SIF)
A standard transfer format established to allow data to be read and used by different computer systems.

Standard positioning services (SPS)
The level of kinematic positioning accuracy that will be provided by **GPS** based on the single frequency C/A code (the normal civilian positioning accuracy obtained).

State plane coordinate system
A grid system that was developed by the National Geodetic Survey for each state. The earth's surface, reduced to sea level, is projected onto a series of plane surfaces. A Lambert conical or Transverse Mercator projection is used, depending on the state's shape (*see* **map projection**). A state can have more than one zone, and each zone has an origin for a grid system. The location of points is expressed in terms of coordinates x and y from this origin.

Statement of qualifications (SOQ)
Documented response by companies wishing to participate in a project as provider in response to an **RFI**.

Stereocomparator
Stereoplotter device that makes measurements from pairs of photographs.

Stereocompilation
Production of a map or chart from aerial photographs and geodetic control data by means of photogrammetric instruments.

Stereodigitizer
Precision stereocompilation equipment for the direct capture of data (i.e., planimetric, coordinate, and elevation) from aerial photographs.

Stereodigitizing
Direct data capture from aerial photographs using precision stereocompilation equipment to collect planimetric, coordinate, and elevation data.

Surface modeling
Analytical techniques to represent three-dimensional surfaces, such as topography.

Symbols
A configuration of lines and curves representing a graphic element.

Symbol recognition
Process of locating and identifying symbols on a scanned map.

Tag image file format (TIFF)
Common industry-standard file format for storing raster data.

Tangible benefits
Measurable benefit.

Target formats
Format that GIS data will ultimately be stored in.

Target GIS
Software that will ultimately use the GIS data.

Target platforms
Hardware that software and data will ultimately be used on.

Targeting
See **paneling**.

Thematic maps
A map made for the sole purpose of communicating a theme or showing statistical information. Commonly found in atlases, there are several types of thematic maps, including dot maps, graduated circle maps, and choropleth (shaded) maps. An example of a thematic map is a shaded map of population density.

Theodolite
A surveyor's instrument for measuring horizontal and vertical angles.

Third-order work
The lowest order of control surveys for which monumentation is authorized.

Three-dimensional (3D)
(1) Representation in graphic form of objects having length, width, and depth. (2) The ability of a computer graphics system to create the geometric representation of an object in the z dimension as well as the x and y.

Tic marks
A set of marks on a map that has known coordinates. Used for registering separate map sheets and as reference points in digitizing.

TIFF
See **tag image file format**.

TIGER
See **topologically integrated geographic encoding and referencing system**.

Tilt
The angle between the optical axis of the camera and a plumb line.

TIN
See **triangulated irregular network**.

Tolerance
The allowable deviation from a standard; the range of variation permitted in maintaining a specified dimension (especially related to GIS data).

Topography
The configuration and location of features on the earth including hydrology, man-made features, and relief.

Topologically integrated geographic encoding and referencing (TIGER)
U.S. Census Bureau reference files developed for the 1990 Census. Similar to the **DIME** files, TIGER was developed as a tool for assigning census data collected by addresses to various geographic areas such as tracts and blocks. Unlike the DIME files, which covered only urbanized areas, TIGER files cover the entire United States. Also, the DIME files were a stand-alone representation of street segments. The TIGER system is designed to fully integrate with actual base map data from the **USGS** 1:100,000 **DLG** files.

Topology
The aspect of a geographic information system which allows specialized analytical operations of spatial search and overlay. Refers to the spatial relationships between the points and lines that define geographic features. Line features are bounded by points called **nodes**. Nodes and **lines** define **polygons**. With topology, a **GIS** understands spatial relationships, such as whether or not a point from one layer falls within a given polygon (e.g., counting manholes in a given sewer district), whether a polygon is to the left or right of a given line (e.g., determining a census tract for a particular address), or whether a polygon has an island polygon within its bounds (e.g., determining whether a special district contains parcels with certain characteristics).

Toponym
A place name or descriptive name.

Translation
The process of transforming data or information from one data format to another. *See* **data translation**.

Traversing triangulation
Method of extending horizontal position on the surface of the earth by measuring the angles and included sides of triangles.

Triangulated irregular network (TIN)
A method of representing a three-dimensional surface using triangular facets, each having an equal elevation. Representation of a surface in this method allows three-dimensional data to be processed with polygon processing.

Trilateration
A method of surveying wherein the lengths of the triangle sides are measured, usually electronically.

U.S. Geological Survey (USGS)
A government agency formed to develop, recommend, and maintain earth science data standards for the federal government; a source of topographic maps of the United States and its outlying areas (both paper and digital).

Undershoot
A graphic line that does not extend far enough to properly meet or intersect another graphic feature.

Universal Transverse Mercator (UTM)
A common rectangular (**Cartesian**) coordinate system based on projection of locations on the earth onto a cylindrical surface. Coordinates are usually expressed in meters north (**northings**) and meters east (**eastings**) from reference axes that define a given zone.

Urban and Regional Information Systems Association (URISA)
A professional organization for individuals concerned with the effective use of information systems technology in the public and private sectors.

URISA
See **Urban and Regional Information Systems Association.**

Users
Those individuals within an organization who use and rely on maps, drawings, and records. Also, those individuals who physically operate computer systems in the course of performing their jobs.

USGS
See **U.S. Geological Survey.**

UTM
See **Universal Transverse Mercator.**

Validation
The process of comparing data converted to a GIS to its source for correctness and connectivity.

Variable scale
The ability of a computer graphics system to generate output in graphic form at a variety of map scales, as specified by the user.

Vector data
In GIS, data comprised of x,y coordinate representations of locations on the earth, taking the form of single *points*, strings of points (*lines*), or closed lines (*polygons*).

Vector GIS
A GIS that represents geographic features using coordinates of **points**, **lines**, and **polygons**. The end product appears similar to the map itself (vs. grid-cell or **raster GIS**).

Vectorizing
Process of converting raster data into vectors.

Vellum
A fine-grained, unsplit calfskin, lambskin, or kidskin prepared paper used for map production.

Vertical control datum
Any level surface used as a reference from which to reckon elevations, usually mean sea level.

Vertical control points
See **control points.**

Voice data entry
The use of voice to record data into a computer.

Wall maps
A map that is so large it can only be handled by pinning onto a wall.

Warping
(1) Photographic distortions or orientation errors. (2) The process by which one set of spatial data is forced to fit another set.

WGS-84
See **World Geodetic System 1984.**

Windows
A computer graphics function that allows selection of a portion of an image for viewing. Usually used in conjunction with **zoom** functions to create a blow-up of a portion of a graphics image.

Work orders
Accounting control device for the design, procurement, and tracking of predominantly field work activity. Depending on the organization, can have widespread influence on facilities tracking documentation.

Work order sketches
See **workprint generation.**

Workprint generation
The production of workprints as part of the work order process that depict the state of facilities within an area as either proposed additions or removals. Generally contain title blocks and other company information and are tracked from preposting through culmination of the construction process to final posting.

Workstations
Single-user computer station designed for high performance graphic and numerical computations. Also referred to as an engineering workstation.

World Geodetic System 1984 (WGS-84)
A unified world datum based on a combination of all available astrogeodetic, gravimetric, and satellite tracking observations. The system is revised as new data materials change the currently accepted values.

WORM
See **Write Once Read Many**.

Write Once Read Many (WORM)
Nonerasable, removable storage device.

Zoom
A computer function to proportionately enlarge or decrease the size of the displayed entities by rescaling.

Index

Absolute orientation, 247
Acceptance criteria, 247
Acceptance tests, 247
Access time, 247
Accuracy (absolute, relative), 247
Acetate overlays, 247
Address matching, 247
Addresses, 247
Admatches, 248
Administrative data, 99
Aerial mapping firms, 186-88
Aerial photographs, 84-85. *See also* Photogrammetry
    color vs. black and white, 87
    defined, 248
    features captured, 87
    reasons for using, 85-86
    scale, 86-87
    overlap techniques, 116
    and raster images, 86, 88
    triangulation, 117
    and vectors, 86, 88
Aerial triangulations, 248
Affine transformations, 248
Algorithms, 248
Aliases, 248
Alphanumeric databases, 248
Alphanumeric information. *See* Keyboard entry
AM. *See* AM/FM systems; Automated mapping
AM/FM International, 193, 248
AM/FM systems, 10
    blending with GIS, 3
    defined, 250
    early line orientation, 3
    and GIS, 12
    origins, 2-3
    and public utilities, 12
AM/FM/GIS. *See* Geographic information systems
American Society for Photogrammetry and Remote Sensing (ASPRS), 248
Analog stereoplotters, 248
Analogs, 248
Analytical stereoplotters, 248
Analytical triangulation, 249
Annotations, 249
Apparent centerline, 249
Applications, 28, 249
Applications developments, 249
Areas, 154
Artificial intelligence, 53
As-built drawings, 249

Aspect, 249
ASPRS. *See* American Society for Photogrammetry and Remote Sensing
Attributes, 249
Automated conversion, 249
Automated mapping (AM), 8-9. *See also* AM/FM systems
    defined, 249
Automated mapping/facilities management. *See* AM/FM systems
Axis, 250
Azimuth, 250

Backlighting, 250
Base maps, 250
Baseline, 250
Batch process, 250
Benchmark tests, 250
Benchmarks, 250
Bit maps, 250
Bits, 250
Blocks, 251
Boolean operations, 251
Breaklines, 251
Buffer zones, 251
Buffers, 11, 251
Bulk loading, 110, 251
Bytes, 251

Cable ripple, 22
Cable throw, 251
CAD. *See* Computer-aided drafting and design
Cadastral data, 98
Cadastral maps, 251
Cadastral property overlays, 251
Cadastral surveys, 251
Cadastre, 252
CADD. *See* Computer-aided drafting and design
Cards and records, 88-89
Cartesian coordinates, 252
Cartographic accuracy, 252
Cartographic quality. *See* Maps
Cartography, 252
CASE. *See* Computer-aided software engineering
Cathode-ray tubes, 252
CCITT Group IV, 252
CD-ROMs, 61, 253
Cells, 252
Census blocks, 252
Census Bureau. *See* U.S. Bureau of Census
Census tracts, 252

Central processing unit (CPU), 252
Centroids, 253
Chains, 154
Change requests, 242-43
Characters, 253
Choropleth maps, 253
Client load, 253
Client-server computing, 51
COGO. *See* Coordinate geometry
Colinearity equations, 253
Command tablets, 55
Compact-disk, read-only memory. *See* CD-ROMs
Compilation scale, 253
Completeness, 253
Compression, 121-22
Computer screens, 47. *See also* Screen copy devices
Computer-aided drafting and design (CAD, CADD), 5-6, 253
 and AM/FM systems, 10
 in GIS data conversion, 93
 and hardware development in 1980s, 49-50
Computer-aided engineering (CAE), 253
Computer-aided software engineering (CASE), 149
Computers, 49, 51
 computer FAX, 58
 mainframe, 64
 minicomputers, 63
 pen-based, 56, 66
Connectivity, 253
Continuous map, 254
Contour intervals, 254
Contouring, 254
Contours, 254
Contractors. *See* GIS data conversion contractors
Control data, 99
Control densification, 126
Control networks, 254
Control points, 254
Conversion, 254
Conversion services firms, 254. *See also* GIS data conversion contractors
Conversion specifications, 254
Coordinate geometry (COGO), 254
Coordinate systems, 254
Coordinate transformation, 254
Coordinates, 254
CPU. *See* Central processing unit
Creators, 255
Credibility, 255
Cursors, 255

Dangle, 255
Data, 255

Data conversion, 1-2, 27-28, 105. *See also* Data conversion planning; GIS data conversion contractors; Hardware
 automated, 123-24
 batch vs. on-line, 124
 and broader data management needs, 45
 and CAD, 93
 contractors and consultants, 31, 38-39, 41-42, 48, 53
 costs and effort involved, 31-32, 196-97
 data-driven approaches, 210
 data sources, 35-37
 developing approach, 32-34
 disruptions caused by, 39-40
 effects on database design, 151-53
 executive committee, 40
 field inventory, 128-30
 field survey, 125-28
 and Global Positioning Systems, 220
 in-house vs. external approach, 53, 184-85, 229-30
 incremental, 51
 information services department, 44-45
 information systems department, 42
 issues to consider, 28-31
 keyboard entry, 109-13
 lines and curves, 123-24
 logistical considerations, 43
 map digitizing, 105-9
 photogrammetry, 113-19
 prioritization, 42-43
 process-oriented strategies, 210
 project management and responsibilities, 40-43
 risk reduction, 44
 scanning, 119-22
 source data matrix, 37, 38
 success factors 44, 206
 symbols, 124
 technical staff, 42
 text, 124
 translation software, 130
 workstations, 58-60
Data conversion planning, 221
 budget, 240
 change requests, 242-43
 communication lines, 237-38
 conceptual database design, 227-28
 contracting flowchart, 232
 contracting outside services, 233
 converted packet size, 238-39
 data maintenance, 243-44
 database design factors, 222-25
 existing data sources, 225-26
 identifying information sources to be converted, 222
 in-house activities, 241
 internal vs. external conversion, 229-30

new data sources, 226-27
physical database design, 228-29
pilot project, 244-45
political factors, 225
positional accuracy, 222-23
production work plan, 235-41
purchasing data conversion services, 230-34
quality assurance, 241-42
reporting requirements, 240-41
requests for information (RFI) from contractors, 230
requests for proposals (RFP) from contractors, 230-31, 233-34
schedule, 239-40
source preparation and scrub, 234-35, 238
Data-driven approaches, 211-13
Data elements, 255
Data encoding, 255
Data entry, 255
Data fields, 255
Data formats, 255
Data layers, 255
Data migration, 256
Data quality, 131, 256
  automated verification, 143-44
  cartographic, 131-34
  informational, 134-36
  manual verification, 144-46
  positional accuracy, 136-43
  verifying, 143-46
Data records, 256
Data segmentations, 256
Data sets, 256
Data sources, 27-28, 30, 35-36, 67-68. *See also* Aerial photographs; Cards and records; Databases; Drawings; Maps
  accuracy, 68
  completeness, 69
  condition, 71
  convenience, 71
  correctness, 70
  coverage, 68-69
  credibility, 70
  defined, 256
  maintainability, 72
  precedence, 72
  readability, 71-72
  reliability, 71
  timeliness, 69-70
  validity, 70
Data translation, 256
Data vintages, 256
Database design, 147
  and application requirements, 151
  area topology, 163
  and attributes, 156-57
  attribute topology, 164
  and available data, 151-52

color use, 160
and computer-aided software engineering, 149
conceptual stage, 148-49, 227-28, 256
and conversion costs, 151-52
and conversion schedule, 153
and data conversion, 222-25
and data relationships, 157-58
elements of, 154, 155
and existing digital data, 158
feature-attribute relationships, 157
feature class relationships, 157
flat files, 164
and future data formats, 158-59
future expansion needs, 153
and geometric integrity, 160-61
graphic elements, 156
graphic structure, 159-62
hierarchical structure, 164-65
implementation stage, 149
interface needs, 166
layers, 161
line topology, 163
logic elements, 154
and maintenance of database, 153-54
node topology, 163
one-to-many relationships, 157-58
physical stage, 149, 228-29, 256
raster image data, 158
relational structure, 165-66
stages, 147-49
symbology, 159
tabular structures, 164-66
text annotation, 161
topology, 162-64
user needs, 150
visibility rules, 162
Database management systems (DBMS), 6-8, 256
  hierarchical, 6-7
  relational, 7-8
Database structures, 256
Databases, 256. *See also* Database design; Database management systems; Multi-participant databases
  accounting-oriented data, 90-91
  adapting existing ones for GIS, 89-90
  assessment-oriented data, 91
  of attributes, 110-11
  commercial, 91-92
  community-oriented data, 90
  completeness, 134
  correctness of data, 135
  data relationships, 35
  design, 6
  developing, 27
  enabling, 210
  engineering-oriented data, 91

flat files, 164
of graphics, 111
and graphics applications, 4
hierarchical structure, 164-65, 256
information sources, 27-28, 30, 35-36
integrity of, 135-36
jointly developed. *See* Multiparticipant databases
and maps, 4, 11
populating, 27, 29, 30, 37-39
relational structure, 165-66, 257
seamless, 3, 275
shared. *See* Multiparticipant databases
tabular, 4, 7-8, 164-66
timeliness, 135
topdown design, 34
types of information contained in, 14-17
Datum, 257
DBMS. *See* Database management systems
Defense Mapping Agency (DMA), 18
DEM. *See* Digital elevation models
Densification, 257
Department of Agriculture. *See* U.S. Department of Agriculture
Department of Defense. *See* U.S. Department of Defense
Destination systems, 257
Diapositives, 257
Differential leveling, 257
Differential positioning, 128
Digital data, 257
Digital elevation models (DEM), 257
Digital holography, 257
Digital image processing, 257
Digital line graphs (DLG), 21, 91, 92, 258
Digital orthophotographs, 258
Digital terrain models (DTM), 258
Digitize (defined), 258
Digitize, manually (defined), 258
Digitizers, 258
Digitizing. *See* Map digitizing
Digitizing tables, 258
Digitizing tablets, 47, 56, 258
DIME. *See* Dual-independent map encoding
Dimensioning, 258
Displacement, 258
Displayable attributes, 258
Displays, 259
DLG. *See* Digital line graphs
DMA. *See* Defense Mapping Agency
Document imaging, 259
Document sets, 259
Dots per inch (dpi), 120, 259
Downloading, 259
dpi. *See* Dots per inch
Drawing layers, 259
Drawings, 259
media, 84

production tools, 84
scale, 83
sheet size, 83-84
symbology, 84
users, 82-83
DTM. *See* Digital terrain models
Dual-independent map encoding (DIME), 259
Dynamic segmentation, 259

Eastings, 260
Edge matching, 260
EDM. *See* Electronic document management
Electric utilities, 21
Electronic document management (EDM), 122, 260
Electrostatic edit plots, 260
Elements, 260
Ellipsoids, 260
Emerging technologies, 209-10
Enabling databases, 210
Encoders, 260
Endlap, 116
Entities, 260
Erasable disks, 260
External conversion, 260

Facilities, 260
Facilities data, 95, 99-102
and land base data, 95, 96, 102-3
Facilities management (FM), 9-10. *See also* AM/FM systems
Facility data, 260
Facility models, 260
FAX, 58
Feature attributes, 261
Features, 154, 261
Federal Geodetic Control Committee, 126, 261
Federal government, 18-19
Fiducial marks, 261
Field data entry stations, 66
Field inventory, 128, 218-20, 261
data collection, 129-30
data verification, 130
Field survey, 125
accuracy levels, 127
conventional, 125-26
horizontal, 125, 126, 127
vertical, 125, 126, 127
Files, 261
Film flattening, 261
Financial services companies, 24
First-order work, 261
Flatbed plotters, 261
Flatbed scanners, 261
FM. *See* AM/FM systems; Facilities management
FMC. *See* Forward motion compensation

Foot (English), 261
Forest Service. *See* U.S. Forest Service
Formats, 261
Forward motion compensation (FMC), 262

Gas utilities, 22
GBF/DIME. *See* Geographic Base File/Dual Independent Map Encoding
GDBS. *See* Geofacilities Data Base Support system
Geocoding, 212, 220, 262
Geofacilities Data Base Support (GDBS) system, 2
Geographic Base File/Dual Independent Map Encoding (GBF/DIME), 91
Geographic coordinate system, 262
Geographic data, 262
Geographic information systems (GIS), 1, 11-12. *See also* AM/FM systems; Data conversion; Databases
  and AM/FM, 12
  area orientation, 3, 11
  blending with AM/FM systems, 3
  buffers, 11
  data-driven approaches, 211-13
  data layers, 11, 95
  data requirements, 28-29, 34-35
  and electric utilities, 21
  federal government uses, 18-19
  and financial services companies, 24
  functions of, 28-29
  and gas utilities, 22
  and governmental users, 12
  local government uses, 20-21
  and insurance companies, 24
  market sectors, 17-18
  and mining companies, 23
  and oil companies, 23-24
  origins of, 2-3
  and private sector, 18, 23-26
  and public sector, 17, 18-21
  qualitative benefits, 13-14
  quantitative benefits, 12-13
  and real estate companies, 25
  and regulated (utilities) sector, 17, 21-22
  and retail companies, 25-26
  seamless database capabilities, 3
  source data, 35-37
  state government uses, 19-20
  technology, 262
  and telephone companies, 22
  and timber and forest product companies, 24
  and transportation companies, 25
  vendor-specific formats, 35
  vendors, 190
Geographic records, 262
Geo-location, 262
Geological Survey. *See* U.S. Geological Survey
Gigabytes, 262
GIS. *See* Geographic information systems
GIS data conversion contractors
  additional services, 201
  aerial mapping firms, 186-88
  automated, 189-90
  bid process, 205-6
  commitment to GIS, 193
  company history, 191
  company resources, 193
  competitive bids, 200
  cost factors, 196-97
  data quality, 197, 203
  data representation, 199
  data translation experience, 203
  database complexity, 201
  database design, 198
  deciding to use, 184-85
  deliverable products, 198-99
  diversity among firms, 185
  engineering firms, 188-89
  evolution of industry, 181-83
  experience, 193-94, 202, 205
  financial reports, 202
  inadequate software, 203
  inadequate workstation capacity, 204
  key personnel, 202
  labor costs, 199
  location, 192, 201
  management ability, 203
  map digitizing firms, 186
  multiple contractors, 197
  offshore labor, 199, 207
  organizational structure, 192
  ownership, 192-93
  pending litigation, 205
  postconversion processing, 200
  price, 196
  quality assurance, 199-200
  range of services, 192
  revenue base, 193
  risk factors, 202-5
  schedule, 200, 201-2, 204
  selection criteria, 191-202
  services, 183
  shifting scrub burden to client, 207
  source document scrub, 198
  source documents, 198
  specification exceptions, 201
  staff acquisition problems, 204
  success characteristics, 208
  target GIS, 197
  technical considerations, 191
  technical plan of operation, 194-96
  using clients' facilities at night, 202
Global Positioning Systems (GPS), 66, 263

controlled image capture, 129
and data conversion, 220
surveys, 126-28
GPS. *See* Global Positioning Systems
Graphic displays, 263
Graphic editing, 263
Graphic elements, 263
Graphic input devices, 55
Graphic records, 263
Graphics
database-generated, 211-13
and databases, 4
databases of, 111
Graphics tablets, 263
Grids, 263
Ground control, 263

Hardware. *See also* Command tablets; Digitizing tablets; FAX; Graphic input devices; Keyboards; Pen-based computers; Pen plotters; Printers; Processors; Raster plotters; Scanners; Screen copy devices; Storage devices; Workstations
development of, 47, 49-53
influence of available technology, 48
influence of data conversion needs, 48
influence of economics, 48
influence of industry experience, 48-49
input, 55-56
in 1980s, 49-51, 52, 54
in 1990s, 51-53, 54
output, 56-58
pre-1980, 49
Harvard University
and GIS origins, 2
Heads-up digitizing, 51, 121, 205, 215-16, 217, 263
Holography, 207
Horizontal control datum, 263
Horizontal control points, 263
Hybrid files, 217
Hydrography, 263
Hydrological (defined), 263
Hypsographic data, 98
Hypsography, 264

IBM Corporation, 2
IGES. *See* Initial graphics exchange standards
Image interpretation (correlation), 264
Image processing, 264
Impact printers, 264
Incremental conversion, 264
Initial graphics exchange standards (IGES), 264
Insurance companies, 24
Integrity, 264
Intelligent infrastructure, 264
Interactive (defined), 264
Interactive graphics, 264

Interior orientation, 265
Internal conversion, 265
Interpretive photogrammetry, 265

Key entry, 265
Keyboard entry, 109
attribute databases, 110-11
cadastral applications, 113
engineering applications, 112
graphics databases, 111
precise coordinates, 111-13
Keyboards, 55

Land base data, 95, 97, 100
administrative, 99
cadastral, 98, 100
control, 99
and facilities data, 95, 96, 102-3
hypsographic, 98
planimetric, 97, 100
Land bases, 265
Land positional accuracy. *See* Positional accuracy
Land-related information systems (LRIS), 265
Landsat, 265
Large format digital holography, 207
Large scale (defined), 265
Layers and layering, 11, 95, 265
and database design, 161
Leveling, 265
Light pens, 265
Line snapping, 266
Linear features, 266
Linen, 266
Lines, 265
Links, 154
Local governments, 20-21
Locations, 266
Lots, 266
LRIS. *See* Land-related information systems

Macros, 266
Magnetic disks, 60-61, 266
Magnetic tape, 62
Mainframes, 266
Management information systems (MIS), 266
Manholes, 266
Map digitizing, 105-6
advantages and disadvantages, 109
equipment, 106-7
process, 108-9
registration, 106
setup, 107-8
workstation configuration, 107
Map features, 266
Map projections, 267
Map scales, 267

Maps, 72. *See also* Aerial mapping firms;
   Automated mapping; Map digitizing;
   National map accuracy standards
   absolute accuracy, 132-33
   contents, 73, 79-80
   and databases, 4, 11
   defined, 266
   digitizing contractors, 186
   electric utility (sample), 77
   graphic quality, 133-34
   incremental conversion, 217
   media, 79
   municipal planning (sample), 76
   production tools, 80-82
   relative accuracy, 132
   scale, 74-75
   sheet size, 75
   symbology, 80
   traditional, 4
   types, 73
   users, 74
   uses, 74
Megabytes, 267
Menu-driven (defined), 267
Menus, 267
Metric photogrammetry, 267
Minicomputers, 267
Mining companies, 23
MIS. *See* Management information systems
Models, 267
Monocomparators, 267
Monumentation, 267
Multiparticipant databases, 167-69
   agreements, 171-72
   content agreements, 178
   executive steering committee, 170
   feasibility studies, 178
   funding, 172-74
   liability issues, 174-76
   maintenance of, 178-79
   ownership, 174
   project champion, 170
   project manager, 171
   risks, 176-77
   standards, 169
   technical project team, 171
Multiparticipant GIS projects, 267
Mylar, 267, 79, 80

Nadir, 267
National map accuracy standards, 114, 115, 136, 139, 268
NAVSTAR satellites, 126-28, 268
Neat lines, 268
Network pair connectivity, 268
Network topology, 268
Networks, 64-65, 268
1929 North American Vertical Datum, 116

Nodes, 154, 268
Nongraphic tabular data, 268
Normalize (defined), 268
North American Datum of 1927, 99
North American Datum of 1983, 99, 116, 268
Northings, 268

Object-based approach, 95
OCR. *See* Optical character recognition
Odyssey, 2
Off-line (defined), 269
Off-line text entry, 269
Offset line, 269
Oil companies, 23-24
On-line (defined), 269
Optical character recognition (OCR), 124, 269
Optical disks, 61
Orientation, 269
Orthophotographs (orthophotos), 86, 118-19, 269
   digital, 119, 122
OSP. *See* Outside plant
Output scale, 269
Outside plant (OSP), 269
Overlaps, 269
Overlays, 269
Oversheets, 269
Overshoot, 269

Pan, 270
Paneling, 270
Parallax, 270
Parcel mapping, 270
Parcels, 270
Pen plotters, 56-57
Pen-based computers, 56, 66
Personal computers, 270
Photogrammetric devices, 65. *See also* Aerial photographs
Photogrammetric stereoplotters, 114, 117
Photogrammetry, 113-14, 270. *See also* Aerial mapping firms; Aerial photographs
   advancements, 119
   aerial triangulation, 117
   and control densification, 126
   digital orthophotography, 119
   digitizing stereomodels, 117-18
   GPS-controlled image capture, 119
   ground control, 116
   orthophotographs, 86, 118-19
   overlap techniques, 116
   process, 114, 115
   reasons for using, 114-16
   soft-copy, 119
   stereoplotters, 114, 117
Photograph interpretation, 270
Photographs. *See* Aerial photographs
Pilot projects, 270

Pixels, 120, 270
Planimetric data, 97
Planimetric map, 270
Planimetry, 270
Plotters. *See* Pen plotters; Raster plotters
Plotting, 271
Point-in-polygon operations, 271
Pointers, 162
Points, 271
Pole cards, 271
Polygon intersection, 271
Polygon processing, 271
Polygon retrieval, 271
Polygons, 154, 271
Positional accuracy, 18, 19, 20, 21, 22, 24, 25, 136, 271
    accuracy requirements, 136, 137, 140-41
    cost issues, 141-43
    and data conversion, 222-23
    measuring (statistically), 139-40
    and scale, 136-38, 140
PPS. *See* Precise positioning services
Precise calculations, 271
Precise positioning services (PPS), 272
Precision, 272
Precision placement, 272
Premarking, 87, 272
Primary sources, 272
Printers, 58
Private sector, 18, 23-26. *See also* Public sector
Processors, 62-64
Professional organizations, 193
Profiles, 272
Public sector, 17, 18-21. *See also* Private sector; Public utilities
Public Service Company of Colorado, 2
Public utilities
    and AM/FM, 12
    and GIS, 17, 21-22
Publication scales, 272
Pucks, 272
Pugs, 272

Quadrangles, 272
Qualitative benefits, 272
Quality controls, 272
Quantitative benefits, 272

Radial displacements, 272
Raster data, 273
Raster GIS, 273
Raster images and format, 57, 86, 88, 120, 133-34, 205, 209, 216
    data compression, 121-22
    and database design, 158
    for land base, 216
    maps, 217
    reference or detail drawings, 217
    using raster data, 216
Raster plotters, 57
Raster-to-vector, 273
Raw data, 95
RDBMS. *See* Databases, relational structure
Real estate companies, 25
Records, 273
Records posting, 273
Reduced Instruction Set Computing. *See* RISC
Reference base, 96
Reflective scanners, 273
Registration, 273
Regulated sector. *See* Public utilities
Relational databases, 273
Relative accuracy, 273
Relative orientation, 273
Relative positioning, 273
Relief displacement, 118
Remotely sensed data, 273
Requests for information (RFI), 230, 274
Requests for proposals (RFP), 230-31, 233-34, 274
Requirements, 29
Resolution, 274
Retail companies, 25-26
RFI. *See* Requests for information
RFP. *See* Requests for proposals
RISC, 51, 62
Rotating drum scanners, 274
Rubbersheeting, 274
Rule-based systems, 274

Satellite imagery, 274
Satellite Probatoire pour l'Observation de la Terre (SPOT), 274
Scales, 274
Scan lines, 274
Scanners, 51, 56
Scanning, 119-20, 189-90, 206-7, 275
    fixing documents in bad condition, 121
    fully automated, 213-15
    and image storage, 121-22
    interactive vectorization, 215-16
    process, 121
    reflective, 119
    and scale standardization, 121
    scanner types, 119
    text documents, 121, 122
    transmissive, 119-20
    uses for images, 122
Schematic drawing, 275
Screen copy devices, 57
Scribe coat, 275
Scrub (defined), 275
Scrubbing, 37
    by client, 207
    and contractors, 198

and data conversion planning, 234-35
Seamless databases, 275
Section corner, 275
Servers, 63. *See also* Client-server computing
Sidelap, 116, 275
SIF. *See* Standard interchange format
Slope maps, 275
Small scale, 275
Software
   advances in, 207
   computer-aided software engineering, 149
   contractors' inadequate possession of, 203
SOQ. *See* Statement of qualifications
Source, 67. *See also* Data sources
Spatial analysis, 275
Spatial data, 275
Spatial topology, 276
SPOT. *See* Satellite Probatoire pour l'Observation de la Terre
Spot elevation, 276
Spot height, 276
SPS. *See* Standard positioning services
SQL. *See* Standard Query Language
Standard interchange format (SIF), 276
Standard positioning services (SPS), 276
Standard Query Language (SQL), 51
State governments, 19-20
State plane coordinate system, 276
Statement of qualifications (SOQ), 276
Stereodigitizers, 276
Stereodigitizing, 276
Stereomodels, 114
   digitizing, 117-18
Sterocomparators, 276
Sterocompilation, 276
Storage devices, 60
   jukeboxes, 60
   magnetic disk, 60-61
   magnetic tape, 62
   optical disk, 61
Surface modeling, 276
Survey input devices, 65-66
SYMAP, 2
Symbol recognition, 277
Symbology, 80, 84
Symbols, 277

Tag image file format (TIFF), 277
Tangible benefits, 277
Target formats, 277
TCP/IP network protocol, 51
Technologies. *See* Emerging technologies; Scanning; Software
Telephone companies, 22
Thematic maps, 277
Theodolite, 277
Third-order work, 277
Three-dimensional (defined), 277
Tic marks, 277
TIFF. *See* Tag image file format
TIGER. *See* Topological integrated geographic encoding and referencing
Tilt, 278
Timber and forest product companies, 24
TIN. *See* Triangulated irregular network
Tolerance, 278
Topography, 278
Topological integrated geographic encoding and referencing (TIGER), 21, 91, 92, 278
Topology, 278
Toponym, 278
Total stations, 65-66
Translation, 278
Transportation companies, 25
Traversing trinagulation, 278
Triangulated irregular network (TIN), 278
Trilateration, 279

U.S. Bureau of Census (USBC), 18, 91
   Geographic Base File/Dual Independent Map Encoding, 91
U.S. Department of Agriculture, 18
U.S. Department of Defense, 18. *See also* Defense Mapping Agency
U.S. Forest Service (USFS), 18, 19
U.S. Geological Survey (USGS), 18-19, 91, 116, 125, 279
Undershoot, 279
Universal Transverse Mercator (UTM) system, 125, 133, 279
UNIX, 51
Urban and Regional Information Systems Association (URISA), 193, 279
URISA. *See* Urban and Regional Information Systems Association
USBC. *See* U.S. Bureau of Census
Users, 279
USFS. *See* U.S. Forest Service
USGS. *See* U.S. Geological Survey
UTM system. *See* Universal Transverse Mercator system

Validation, 279
Variable scale, 279
Vector data, 279
Vector format, 121, 123, 209
Vector GIS, 280
Vectorization, 217
   automated, 213-15
   interactive, 215-16
Vectorizing, 280
Vellum, 280
Vertical control datum, 280
Voice data entry, 280

Wall maps, 280
Warping, 280
Windows, 280
Work orders, 280
Workprint generation, 280
Workstations, 58-59, 63, 281
    communications between, 64-65
    digitizing, 59
    review/edit, 59
    review/tabular attribute data input, 59
    X terminals, 53, 60

World Geodetic System 1984, 128, 281
WORM optical disks, 61, 281
Write Once Read Many. *See* WORM optical disks

X terminals, 53, 60
X Window System, 51

Zoom, 281